Offshore Oil and Gas
PEOPLE

Overview of Offshore Drilling Operations

Amanda Barlow

DEDICATION

The author would like to thank the following two people for their ongoing support, inspiration and enthusiasm at the idea of writing this book. It was only with the encouragement of these two amazing women that this project got off the ground. You are both true mentors with hearts of gold to match your professionalism and dedication to the energy industry. Thank you for your sincere and ongoing support through the Pink Petro community.

<u>Katie Mehnert</u> – CEO and Founder of **Pink Petro**

"The global energy community disrupting the gender gap | 120 countries | We elevate energy stories & the talent powering our world"

<u>Tina Peters</u> - Licensed Driller- FL & MS President/Owner - MALLARD, INC. Woman Owned, DBE Certified 30 + year experience tina@mallard-inc.com

CONTENTS

FORWARD

This book was written in an attempt to explain what it's like to work on an offshore oil and gas rig so everyone can have a better understanding of what's involved - beyond what you see in the "Deepwater Horizon" movie. While this movie actually does a very good job of portraying what it's like working on a rig (if you haven't already seen it then go and see it!), there is a lot more to understand about the facility and the people who work on it. I hope you learn something from the book and gain a better understanding of the jobs the many highly specialized professionals and tradespeople perform in these hazardous conditions.

While the oil and gas industry employs hundreds of thousands of people worldwide either directly or indirectly, most of those people work in "downstream" operations and have never experienced living and working on an offshore rig. Through reading this book they will gain a better understanding of the collaboration involved between the oil and gas companies, drilling contractors and third-party service providers to find the world's future hydrocarbon reserves.

CHAPTER 1

OVERVIEW OF OFFSHORE DRILLING OPERATIONS

The drilling of offshore oil and gas wells is almost as far "upstream" as you can possibly go in the oil and gas industry. "Upstream" refers to the source of the supply chain for hydrocarbon products while "downstream" refers to the refining and transporting to the final end user. Naturally there's a whole range of stages between these two end points but the stage that most people have least exposure to is the initial drilling of the wells, as only the specialists who are involved in the operations are allowed access to this environment. Barring the few

VIP dignitaries that visit the rigs, it is practically impossible to visit a rig as a "tourist" unless you are directly involved in the operations. Due to the extremeness of the locations and the hazards involved in not only getting the people to the rig but also guaranteeing their safety while on the facility, only appropriately trained essential personnel are allowed to fly to offshore drilling rigs.

With the majority of oil and gas industry workers being involved mainly in the downstream operations within the industry I thought it would be of interest to people to read about what is involved in the drilling process. It may give you some useful information to pass on to young adult family members or friends who are wanting to start out a career in the oil and gas industry, so they can see what skills are needed to be able to work in the offshore drilling environment. The technical expertise in this industry is phenomenal and for anyone interested in geoscience or engineering disciplines you couldn't ask for a more dynamic and exciting career. As well as exploring all the different professions and trades of people involved in the operations, I'll also explain the hierarchal structure of the work force - from the oil and gas company men to the drilling contractor, third-party contractors and all the other specialists involved in the drilling of a well. It is a huge collaboration by all people involved and everyone has their part to play in the successful drilling of a well.

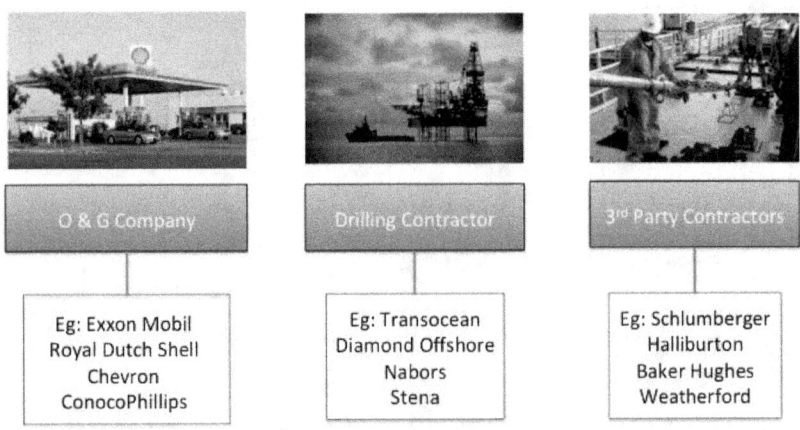

O & G Company	Drilling Contractor	3rd Party Contractors
Eg: Exxon Mobil Royal Dutch Shell Chevron ConocoPhillips	Eg: Transocean Diamond Offshore Nabors Stena	Eg: Schlumberger Halliburton Baker Hughes Weatherford

But before we discuss the people who work on a rig, lets first get a brief introduction to the drilling rigs themselves and the most common types of rigs in use today.

Offshore Drilling Rigs

Offshore drilling rigs fall into two main categories: bottom founded units that have legs that sit on the seabed, and rigs that float on top of the water. They are all commonly referred to as mobile offshore drilling units (MODU's). Deciding what type of rig to use is most commonly dependent on the depth of the water at the well location.

The main types of offshore drilling rigs are:

Jack-Up Rigs

A jack-up rig consists of a platform that is supported by usually three legs (but sometimes four) whose footings are seated on the sea floor and the rig is then "jacked up" to a specific height above the surface of the sea where it won't be adversely affected by wave and tidal movements. The jack-up rig is towed to location with its legs elevated and once on location, the legs are lowered to the seafloor and the platform is "jacked up" above the wave actions by means of hydraulic jacks. Due to the fact that there is a limit on how high a rig can safely be jacked-up, this type of rig can only be used in water depths up to approximately 550 ft (167 m).

Semisubmersible Rigs

A "Semi-sub" is a floating unit that obtains most of its buoyancy from ballasted, watertight pontoons located below the water surface and wave action. With its hull structure submerged at a deep draft, the semi-sub is less affected by wave loadings than a normal ship. Semi-subs are commonly subdivided into generations, depending upon the year they were built and the water depth capability. Generation 1, 2 and 3 rigs commonly use mooring systems and operate in waters less than 1,500 ft

(500 m) while generation 4, 5 and 6 rigs can drill to a water depth of up to approximately 10,000 ft (3,000 m).

Generation	Water depth		Dates
First	about 600 ft	200 m	Early 1960s
Second	about 1000 ft	300 m	1969–1974
Third	about 1500 ft	500 m	Early 1980s
Fourth	about 3000 ft	1000 m	1990s
Fifth	about 7500 ft	2500 m	1998–2004
Sixth	about 10000 ft	3000 m	2005–2010

Semisubmersible rigs are kept on location over the well by a computer-controlled system known as "dynamic positioning" (DP). Position reference sensors, combined with wind sensors, motion sensors and gyrocompasses, provide information to the computer pertaining to the vessel's position and the magnitude and direction of environmental forces affecting its position. This knowledge allows the computer to calculate the required steering angle and thruster output for each thruster. This allows rigs to operate where mooring or anchoring is not feasible due to deep water or seabed problems. Dynamically positioned vessels are categorized into three classes: Class 1 has no redundancy so loss of position may occur in the event of a single fault. Class 2 equipment has redundancy built in to the system so that no single fault in an active system will cause the system to fail. Class 3 systems also have to withstand fire or flood in any one compartment without the system failing. These vessels have at least two independent computer systems with a separate backup system.

Drillships

Drillships have the functional ability of semisubmersible drilling rigs but being a ship means they have greater mobility and can move more quickly under their own propulsion from drill site to drill site. They generally have a higher POB (persons on board) capability, as they require a full marine crew to operate the vessel as well as a drilling crew for drilling operations. Like semisubs, drillships are subdivided into generations, depending upon the year they were built and the water depth capability, and also classes of dynamic positioning capabilities.

Types of Oil and Gas Wells

Wells are classified according to their purpose and fall into three major categories. Exploration wells are tentative ventures that drill in new areas with the hope of discovering untouched resources. Appraisal wells are drilled to evaluate the characteristics of existing hydrocarbon accumulation discoveries and production/development wells are drilled specifically for commercial production of oil and gas from proven hydrocarbon reservoirs. The different types of wells are basically all drilled in the same way but they differ in the way they are completed. In exploration wells there is a very strong focus on end-of-well testing to determine petrophysical properties of any potential reservoir zones, while production wells are completed to allow for future infrastructure to be set up over the well to enable the extraction of the hydrocarbons. Every stage of all types of wells is performed by a collaboration of many teams of people who have very specific expertise in the tasks being undertaken. Experts are flown in from all around the world, which results in a rig full of the most culturally diverse workforce unlike any other workplace in the world. It's this cultural and professional diversity that makes working offshore more exciting and rewarding than any land-based job could ever be.

Working Offshore

The following chapters in "Offshore Oil and Gas PEOPLE" will explore the different professions and trades that offshore workers require to be able to perform the highly specialized jobs that are involved in the successful drilling of offshore oil and gas wells. While it's possible for anyone to find a niche occupation within the many varied roles that are represented offshore, being able to work in this type of environment is another story. Working 12+ hours every day for up to 28 days straight is not for everyone, especially when it might be on the other side of the world from your family and friends.

Before even being allowed on the chopper to fly to the rig, every worker must have completed a sea survival course, which includes helicopter underwater escape training (HUET). This 2-3 day course has to be renewed every four years and is a mandatory prerequisite to be allowed to fly to an offshore facility. The HUET component involves being strapped into a helicopter simulator module, which is submerged in water and then rotated until the occupants are sitting upside down while submerged completely under water. They then have to perform an emergency escape out of the windows of the module and swim to the surface of the pool. Not everyone's idea of fun! This step alone has been known to end some peoples offshore career before it even begins.

While there are many transient workers offshore who come out from time-to-time, the articles to follow will cover only the main personnel who provide the core of the offshore team. There are many others who act as support crew onshore but their positions are beyond the scope of these articles. The offshore workers will be grouped according to the following categories:

Oil and Gas Company

- Permanent company employees – Wellsite Managers, drilling engineers

- Contractor company representatives – Logistics coordinators,

- Wellsite geologists, HS&E personnel, drilling fluids engineers (mud engineers)

Drilling Contractor

- Offshore Installation Manager (OIM)

- Drilling crew – toolpushers, drillers, assistant drillers, derrickmen, roughnecks

- Deck crew – deck pusher, crane operators, roustabouts

- Subsea crew – subsea engineers

- Marine crew – drillship captain, dynamic positioning operators (DPO), ballast control operators (BCO), storemen

- Radio operators

- Medics

- Rig Safety and Training Coordinator

- Mechanical tradespeople
- Electrical tradespeople

Third Party Contractors

- Mudloggers/Data engineers
- Fluids control engineers
- Casing personnel
- Cementing personnel,
- MWD/LWD engineers (Measurement while drilling/Logging while drilling)
- Directional drillers
- Wireline engineers
- Well testing engineers
- ROV operators (remotely operated vehicle)
- Catering contractors – camp boss, cooks and kitchen staff, utilities (cleaners)
- Many various specialists

I will explain what all these roles entail, what qualifications and experience is needed to work in these positions, and also the day-to-day routine that all offshore workers have to follow. There are uncompromising health, Safety and environment (HS&E) standards that must be adhered to and countless meetings and reports are an unwelcomed, but very necessary, part of everyone's responsibilities offshore. Due to the 24-hour operations, men and machinery can be pushed to their limits – and quite often are! The unpredictability of drilling operations means that everyone has to be on their toes and alert for any signs of danger. As the Deepwater Horizon incident has shown, ignoring warning signs can lead to catastrophic events. If something goes wrong offshore, it can go disastrously wrong!

CHAPTER 2

DAILY ROUTINE OFFSHORE

Safety Training

Before anyone can fly to an offshore facility, regardless of whether they work for the oil and gas company, the drilling contractor, or third party contractors, they have to have a current sea survival and helicopter underwater escape training (HUET) certification. And before even being permitted to do this course they have to pass an associated medical examination that checks for any problems that could compromise your health while performing sea survival skills, fire fighting drills and being dunked under water during the HUET. If you are a permanent employee of the company who you represent then the company will generally pay for your certification. If, like me, you are an independent contractor then you have to pay for it out of your own pocket.

There are a few industry-standard survival courses that can be undertaken and they increase in degrees of complexity depending on how extreme the working environment is going to be where you'll be working. Additional safety measures are continually being added to the course as newer and more sophisticated equipment and emergency breathing systems evolve within the industry. The following is a list of the standard courses:

BOSIET – Basic Offshore Safety Induction and Emergency Training.

The BOSIET course is a minimum requirement to work offshore. The two day (generally) course consists of four modules; Safety Induction, Helicopter Safety and Escape (HUET), Sea Survival and First Aid, Fire Fighting and Self Rescue. During the safety induction delegates gain an understanding and awareness of emergency response procedures on offshore installations. The BOSIET is required for cold-water areas and includes additional training in the use of survival suits and emergency breathing systems during the HUET module.

Although the HUET part of the training doesn't sound too arduous, it's surprising how hard it is to hold your breath when you are securely restrained in the module with a 4-point harness and turned upside down underwater. Panic tends to take over which increases the brain's need for oxygen so the body's natural reflex to breath becomes even stronger. Knowing how hard it is to remain calm in a simulated exercise only makes the thought of being in a real-life helicopter ditching scenario all the more terrifying. It's always in the back of your mind when you work offshore but you just hope it never happens to you.

This YouTube video gives an excellent idea of what to expect when you do your BOSIET: https://www.youtube.com/watch?v=oLqXlEGxX xo

TBOSIET – Tropical Basic Offshore Safety Induction and Emergency Training.

The TBOSIET is almost the same as the BOSIET but it is for people who will only be working in tropical waters. This course doesn't involve the use of cold-water survival/transit/submersion suits or the use of emergency breathing systems during the HUET component of the course.

Given the nature of the job, and the fact that all offshore workers could be working in many different environments over the 4-year validity of a training course certification, it's wise to do the BOSIET rather than the TBOSIET because you are then covered for working in both cold and tropical water environments.

FOET – Further Offshore Emergency Training

This course is only applicable to participants who

have a current BOSIET card that is still current within its four years of validation. It is designed to assist in maintaining participant's skills that are required to respond effectively to offshore incidents and offers a practical, theory reduced opportunity to practice how to respond to a helicopter incident and all components of survival at sea including fire fighting, self-rescue and first aid. It is generally a 1-day refresher course but can only be undertaken if your previous certification hasn't expired. If you delay in getting re-certified before the four years is up then you have to do the full BOSIET course again.

Travelling to the Rig

In well-established oil and gas fields there will be a heliport dedicated to flying workers to the offshore facilities. For exploration wells that are being drilled in unestablished fields there may just be a temporary check-in facility at the closest airport to where the rig will be drilling. Many of the people working on the rig will reside in a location other than where this heliport is located and many specialists are based in countries all around the world and make the long journey to the heliport in the days leading up to their crew-change day.

The hitches offshore are generally "even-time" and can be up to four weeks long which sees the workers on the rig for up to four weeks straight and then they fly home for four weeks. On offshore production facilities the rosters are generally a lot shorter than this but drilling rigs tend to have anywhere from two to four week rotations. The cost and convenience to fly people from the other side of the world usually sees these workers working four-week hitches. The closer the workforce is to a home base then generally the shorter the hitches will be. Generally it's wise to pack enough toiletries to last for 4 weeks...just in case!

Once everyone has "checked in" at the heliport on the day of travelling to the rig, they will have their bags either manually checked or placed through an x-ray machine. There are strict rules about what can't be taken offshore, and especially in the cabin of the helicopter. There are total bans on weapons, alcohol and cigarette lighters and mobile phones have to be checked in with your luggage and are not allowed to be taken in the cabin of the chopper. You must always carry a current BOSIET card with you at all times and quite often you are also required to carry a passport. Different countries have different requirements for documentation so you need to check what you will need before leaving home. If you cannot produce these documents upon checking in then you will most probably not be allowed to fly offshore. This then has a knock-on effect because your back-to-back who is on the rig and due to fly home when you get there, will not be allowed to leave the rig.

Once the bags are checked in everyone is required to have a breath alcohol test and sometimes a random (and sometimes not so random!) drug test. You can also expect to have your body physically patted down and also brushed over with a metal-detecting wand. Most companies today have a "zero tolerance" policy in regards to drugs and alcohol and if you fail the tests at check-in then you will not only miss going to the rig for that hitch but will most likely never be allowed on that rig again for the duration of that drilling campaign.

With all the testing out of the way the passengers are directed into a briefing room where they will find a life jacket and set of earmuffs that are to be worn for the flight to the rig. Everyone has to watch a helicopter briefing video before every flight, including when you leave the rig to fly home. The video shows the safety equipment on board the particular aircraft you'll be travelling on and also gives a brief recap of how to escape from the helicopter should it have to land on the water. With the briefing completed everyone is led out to the tarmac in single file and instructed to board the chopper.

The only items you are allowed to carry with you

in the cabin are magazines or soft-covered books. No newspapers, hardcover books or iPods (or any personal electronic devices) are allowed.

Once the chopper is in flight it is extremely noisy in the cabin and it's necessary to not only wear the earmuffs but also earplugs underneath these. Unlike in the "Deepwater Horizon" movie nobody generally talks during the flight because you can't hear a thing above the engine and rotor noise.

If you are working in cold weather environments you can expect to have to wear the following gear for the chopper flight to the rig: 3 layers of clothing, a survival/transit suit, an inflatable life-jacket, emergency breathing apparatus, ear plugs and earmuffs. It's far from being a joy flight!

Arrival on the Rig

When the helicopter lands on the rig the incoming crew disembark and head to the heli lounge in the accommodation block, which is usually close to the helipad. The chopper keeps its engines and rotors running while the incoming bags are offloaded and the outgoing bags are loaded into the cargo hold.

While this is being done the incoming crew gets a brief handover of operations from their outgoing back-to-back while they swap life jackets and earmuffs. This usually takes place in about 10 minutes and as soon as the bags are loaded the outgoing crew are lined up and marched out in single file to the waiting chopper.

The company man and the offshore installation manager (OIM) then brief the incoming crew on the current operations on the rig. Anyone who is new to the rig has to do a full rig induction before they can start work. Everyone else drops their bags off in their room and either starts work if they are on day shift, or attempts to get some sleep if they are going onto night shift.

The standard of the accommodation cabins varies from rig to rig. The older rigs have many 4-man rooms with or without ensuites, and communal shower blocks for those without ensuites. Most third-party contractors and lower rank drill crew workers will be allocated these rooms while the supervisors will usually get allocated 2-man rooms that generally have an ensuite.

The OIM and the day company man (Wellsite Manager) are commonly the only people who have a room to themselves. The 2 and 4-man rooms are generally occupied by a mix of dayshift and nightshift workers to minimize the amount of people in the room at any one time.

Once you leave your room at the start of your shift it is accepted etiquette that you take with you everything you need for your shift and not go back in to your room until your shift is over. By doing this the people who are on the opposite shift and trying to sleep will not get disturbed.

Daily Schedule

The operations are 24-hour and everybody works 12-hour shifts. There are two main shift times and these are "12 to 12" (midday to midnight or midnight to midday) and "6 to 6" (6am to 6pm for dayshift and 6pm to 6am for nightshift). The shifts are commonly referred to as "tours" (pronounced "towers") and generally half an hour before each "tour" starts there is a mandatory "Pre-Tour" meeting that everyone going onto that shift must attend so they can find out what operations have been carried out during the previous 12-hour shift while they have been sleeping. A lot can happen during a 12-hour shift so it's imperative that everyone knows what to expect when they start work. Any issues with personal safety, process safety or just the stage of current operations are discussed.

As with all the meetings offshore, everyone has to sign an attendance sheet as proof of being present. There can be anywhere from 100 to 200 workers onboard the rig at any one time so these meetings can be very busy.

In addition to the pre-tour meetings there will quite often be an additional third-party meeting at about 7am and 7pm. The company man/wellsite manager holds these meetings and only the supervisors need attend. Everyone briefly explains what they will be doing for the shift and any safety concerns are raised and discussed. There can be many concurrent operations being undertaken while the drilling operations proceed and everyone has to be aware of timings and how their tasks will affect all the other operations.

There will be a daily "morning call" at around

0800hrs (depending on time zones between the rig and head office), which is a phone (and sometimes video) conference call between the rig and the drilling superintendent who is based onshore in the head office of the oil and gas company that is drilling the well. This meeting is attended by the dayshift company man/wellsite manager, the OIM, the DLC (logistics coordinator), possibly the wellsite geologist, and any other person who may need to provide specific technical information on the current operations.

Every Sunday (generally) there are mandatory weekly safety meetings that are held at 0100hrs and 1300hrs for all off-tour personnel. There may also be one at 0700hrs or 1900hrs for the personnel who work the 6 to 6 shift. These meetings are held by the rig safety and training coordinator (RSTC) and will cover "safety shares" about incidents that have been reported within the industry in recent weeks.

As well as a weekly safety meeting there is also a weekly fire and abandon rig drill. These are compulsory for all non-essential personnel and the timing of them usually alternates each week so the one shift doesn't keep getting woken up all the time.

If the alarm goes off while you are sleeping then you have to get up and don full personal protective equipment (PPE) and muster out at the lifeboats on the deck. The drill usually takes up to an hour to complete.

As you can see, there are a lot of meetings to be attended when you work offshore and many of them are either before or after your 12-hour shift.

Depending on how busy your back-to-back has been while you've been sleeping, you can also have a lengthy 'handover" time between shifts and together with the pre-tour meeting you can find your shift stretching into 13 hours. When you are working this schedule for up to 28 days straight then you need to be able to manage personal fatigue so you don't compromise your safety, and the safety of everyone else onboard.

In the next chapter I will describe the facilities that can commonly be found on most rigs. With so many people working and living in such a confined space it's important to have a well-structured routine not only out on the deck but also in the accommodation block.

Because commonsense isn't always a given, and with so many different cultures represented, it's necessary to also have stringent rules about how you perform day-to-day tasks that you take for granted at home.

CHAPTER 3

OFFSHORE LIVING CONDITIONS

In this chapter I'll explain what the facilities are like on offshore drilling rigs. The later generations of rigs have very comfortable modern living conditions, albeit somewhat compact compared to most onshore accommodation facilities.

The accommodation block on all rigs is found under, or beside, the helideck. The helideck is the reference point for the locations on the rig with this being the "forward" end of the rig and the other three sides being the starboard, aft and port.

The accommodation block houses not only all the sleeping quarters but also the main offices for the supervisors on the rig.

Atop the accommodation block and next to the helideck there is sophisticated satellite communications equipment, which provides phone and internet connections to the outside world.

The primary lifeboats are always found on the port and starboard sides of the accommodation block and secondary lifeboats at the aft end of the vessel. It is essential that all personnel familiarize themselves with the quickest route to their allocated primary and secondary lifeboats from both their sleeping quarters and also their workplace. Most rigs are a maze of levels and pathways and not knowing the quickest route to the lifeboats can cost you valuable time in an emergency situation. This is one of the reasons why fire and abandon rig drills are performed every week so it becomes second nature for everyone to access their emergency stations.

The top level of the accommodation block houses the "bridge", which is the main office for the marine crew, especially when the vessel is in transit. The radio operator also generally works out of the bridge and is the main point of call for any emergency situations on the rig.

There can be anywhere from two to several levels within the accommodation block and sleeping cabins can be on all levels, but generally the higher you are up the food chain, the higher the level you will sleep on. As mentioned in the previous article, the cabins are generally two-man or four-man rooms.

All later generation vessels will have ensuite bathrooms in each room but there are still many old rigs around the world that have four-man rooms and communal bathrooms.

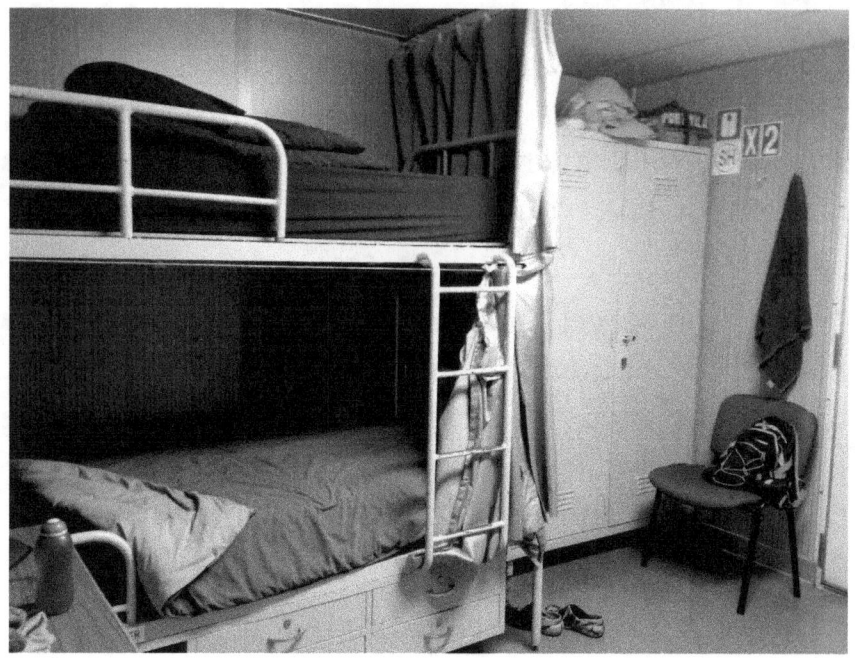

For safety reasons you are not supposed to keep your door locked while in your room because if there is an emergency then people need to get into your room to make sure you have woken up and preparing to evacuate the rig.

During a fire drill it is the utilities job to systematically go from room-to-room checking that no one is left in any of the rooms while the crew are preparing to abandon the rig.

Galley

The galley and dining room/mess are normally found on a lower deck and meals are provided around the clock. The main meal times are at 0500-0700hrs, 1100-1300hrs, 1700-1900hrs and 2300-0100hrs, but the galley is open almost 24 hours a day so you can obtain snacks in between meal times.

The quality of food varies from rig to rig but is usually pretty good. Depending on the cooks, you can sometimes get exceptional quality of meals. It's not unusual to get top restaurant chefs working offshore because they can quite often make much more money as a cook offshore than working in a restaurant in the city.

Many rigs even have pastry chefs who not only keep a supply of yummy cakes and desserts in the fridge but also bake fresh bread daily.

Being a 24-hour operation means that some unlucky people have to work over public holidays. The galley crews always put on a special spread over holidays like Easter and Christmas so even though you have to work a 12-hour shift you still get to have a celebratory dinner.

The festivities are dampened somewhat by the fact that you only have 20-30 minutes to enjoy the dinner before having to go back to work and also, with alcohol being forbidden offshore, there's only non-alcoholic wine or beer to wash down the feast. Being at home with family and friends is always a better option!

Recreational Facilities

Most rigs these days have reasonably well-equipped gymnasiums on board. The bigger the rig, generally the bigger the gym is likely to be. I have been on rigs where the gym is the width of the length of the treadmill and if two people are in there at the same time then it is crowded! Others have had two or more training rooms – one for cardio equipment and another for weight training gear.

Big TV screens on the walls is also pretty commonplace, and someone has usually rigged up some sort of sound system that you can plug your iPod into if you want to listen to music.

As well as training in the gym, many rigs also allow the helideck to be used for recreational activities in between helicopter operations. In sunny climates the helideck is also a favourite place for rig crew to work on their tans...not a wise choice if you're one of the few females working on the rig.... especially given that there are nearly always closed circuit TV cameras tuned into the area 24 hours a day!

Many rigs also have additional recreational activities

like table tennis tables, pinball machines, saunas, and probably lots of other things that I haven't personally seen. It used to be common to have a theatre room where people would watch "videos" but these days most rigs provide TV's and DVD players in all the bedrooms so that has pretty well killed the social interaction that having movie nights used to provide.

Most people also have hard drives full of movies that they take out with them and watch on their laptops while lying in bed.

Communications

Satellite communications equipment is becoming ever-increasingly sophisticated offshore with online communications for workers wanting to connect with loved ones back home now an accepted standard of living conditions offshore.

Naturally though, this is only a side benefit of the oil and gas company's need to transfer ever-increasing amounts of data between the field operations and the head office. Real-time data packages sent during key periods of the drilling operations (such as drilling through the reservoir formations and wireline operations) require reliable and fast speed data transfer rates.

There will always be two separate communication networks; one provided by the drilling contractor and one provided by the operating oil and gas company.

Both of these networks provide free wifi for the workers onboard the rig but it is usually of very limited bandwidth and very slow. It is however, a way for people to stay in touch with family and friends and instant messaging via a mobile device is probably the most popular method of communicating with the outside world while on the rig.

All offshore rigs still have an "old-fashioned" telephone in a booth somewhere so people can make free phone calls home. The commonsense approach prevails of limiting calls if others are waiting to use the phone. Most workers these days travel with a laptop or iPad and access the internet while on the rig, just as they would if they were at home. The only difference is that they have to expect certain sites to be blocked, such as porn sites and gambling sites. These two are big "no-no's" and generally YouTube is inaccessible also as it uses too much data for the overloaded system to handle.

Mobile/cell phone use outside of the accommodation block is absolutely prohibited, as is taking photos. Because of the possibility of gas being encountered during drilling operations, all non-intrinsically safe electronic devices are forbidden. Should photos need to be taken for operational purposes then a special intrinsically safe camera has to be used.

Normal cameras are able to be used but only with the issuing of a "hot work permit" which has to be signed off by several senior people who deem the work area to be safe for the task. The use of cell phones outside of the accommodation block can, and in many cases has, resulted in instant dismissal and an unplanned flight on the next available chopper off the rig.

Medical Facilities

Most rigs will have day shift and night shift medical personnel on board, and usually one of them will be a fully qualified medical doctor. They work out of a reasonably well-equipped "hospital" which is located in the accommodation block and usually in an area that can be easily accessed from the helideck.

With many people working month-long hitches offshore it's essential to have a facility that can cope with not only operational emergencies arising from accidents and incidents on the rig but also everyday medical problems as well. If someone unknowingly brings out a virus that they picked up while on break then it can travel around the rig faster than a wildfire.

With the accommodation block being a watertight structure it is necessary to maintain a comfortable air quality and temperature using ducted air-conditioning. Added to this problem is the fact that many people spend their whole 12-hour shift working in customized shipping containers that are pressurized and have no windows and have to share their air with several other teammates in the very confined conditions. If any of these people have a cold then you can bet they will all eventually get it.

While getting a cold at a job in the city may not be that big a deal – you just call in sick for the day and drive to the nearest pharmacy to get some cold and flu tablets – when you are hundreds of miles out to sea and still have weeks before you will be flying home then it is very inconvenient. Not only do you have to continue to do your job for 12-hours a day (as there are no relief workers out there!) but you can't drive to a shop and pick up any medication.

Gastroenteritis is another common illness that crops up from time to time and is easily spread around the rig by contact with handrails and doorknobs. And when you consider that NOT using the handrails when you are walking up or down the stairs is also a sack-able offence then it can be hard to avoid picking up germs if you aren't diligent with personal hygiene. There's also a chance people can pick up exotic diseases during their transit time to the rig, with many workers flying in from all around the world.

Any sick or injured people who are deemed unable to continue with their duties will be medevac'd off the rig at the earliest convenience. The medics who work on the rigs are generally ex-military personnel and well equipped to deal with the first aid treatment of serious trauma injuries.

Depending on how far offshore and how remote the area is it can take several hours to get to a land-based hospital so you definitely don't want to be hurting yourself out there!

Laundry Facilities

One of my favourite things about working offshore (that only a mother of three kids can appreciate) is that you don't have to do any cooking, cleaning or washing of clothes. It's all done for you! The most you have to do at meal times is scrape your leftovers into a bucket and leave your plates and cutlery on a bench for someone else to wash. How brilliant is that?!

But wait…it gets better. Your bedroom is not only serviced but generally your bed is made for you and the bathroom cleaned. But the best part of all is that before you go to bed, you place your work clothes from that day in a laundry bag and place them on the floor in the corridor outside your cabin and from there they get collected by one of the utilities and washed, dried, folded and placed back on the floor outside your door so by the time you wake up they are ready to wear again. If only life was that easy at home!

<u>Pre-Travel Briefing</u>

When travelling to a rig for the first time, it's

recommended you contact anyone you may know from your company who is already on the rig to ask the following questions:

- **How cold are the living quarters?** Even if the rig is in a tropical region, it pays to take a jacket out with you. The air conditioning can be savage on some rigs.

- **What types of power outlets are on the rig?** As offshore rigs are built and work in different locations all around the world they can have power outlets that differ from what you use at your home base. Generally they are UK or US type outlets but it pays to make sure so you can take any adapters you may need for charging your personal electronic devices.

- **What are the dress standards in the living quarters?** Most rigs will have a "closed-in shoe" policy or "no slip-on shoes" policy. It pays to take closed-in shoes with you just in case. You need to wear closed-in shoes for the chopper ride to the rig anyway. I've seen a person rock up at the heliport wearing only thongs (Jandals) and didn't even have a pair of safety boots with him because he expected to get these on the rig. He was not allowed to fly to the rig.

- **Are you allowed to wear work clothes inside the accommodation block?** Work clothes are generally not allowed to be worn in the accommodation quarters, even if they are brand new or recently washed. Some rigs will allow clean work clothes to be worn so it's best to check so you can be sure. You don't want to be embarrassed by getting chastised about your mistake while standing in line to get your dinner.

- **What areas are set-aside for smokers?** Smoking of cigarettes and ecigarettes is only allowed in designated areas/rooms. These areas can be out on the deck, inside rooms, or both. Lighters will be provided in these areas so there is no need to take them out with you – and it is prohibited to do so.

- **Knowing how you'll be able to contact your loved ones** while you're away is handy to know so your family knows what to expect. The slow Internet speeds can make it difficult to download apps while on the rig so it's best to download any apps you might need on your phone before you leave home.

****

As you can see, there are many rules and regulations

that have to be followed when working offshore, not just when you are out on the deck doing your job but also in the living quarters.

With so many people working and living in such a confined space it's necessary to set the boundaries so everyone knows what's expected of them while on the rig. There can be severe consequences for non-compliance and a free ride on the next chopper off the rig can be your punishment for not following the rules.

Being "run off the rig" is a more common occurrence during the downturn we are experiencing now, when jobs are hard to come by and there are plenty of rule-abiding workers waiting in the wings to replace less dedicated workers. The old "three strikes and you're out" rule has almost become extinct and replaced with a "zero tolerance" policy as health and safety standards become increasingly strict and rigidly enforced.

The "foolhardy" or "larrikin" behavior of yester-year is no longer accepted behavior offshore. While the comraderie amongst some crews is legendary, you will soon find out that life on a rig is far from easy. Many jobs are labor-intensive and 12-hour shifts can be very arduous for the uninitiated.

In Chapter 4 I will start to break down the different

jobs, professions and trades that can be found on offshore rigs. Starting with the drilling contracting company I will explain all the positions, starting with the least experienced workers and building up to the rig manager.

If you've ever wanted to work offshore but haven't really understood what opportunities there are, then continue reading so you can see the range of positions and qualifications needed to get a start offshore.

CHAPTER 4

DRILLING CONTRACTOR – DECK CREW

Now that we've covered the comfy aspects of working offshore…like the free food and gym…it's time to get serious about what people actually do on an oil rig. The reason the "off-tour" facilities are so good is because when you are "on-tour" you work bloody hard! When you are working 12+ hours every day for up to 28 days straight then you don't have much spare time to be doing cooking, cleaning and washing.

When people hear the term "oil rig workers" the mental picture that immediately comes to mind is that of roughnecks throwing tongs and slips around the drill pipe on the drillfloor. Without a doubt, this would be the most dangerous and physically demanding job on an offshore oil rig, and the one the public most commonly associates with the drilling industry.

While the drilling crews are the "face" of the industry they are just a small part of the total workforce that contributes to the successful drilling of each and every well drilled offshore.

In this chapter I'll start breaking down the workforce and explaining all the different roles that are performed offshore. Starting with the drilling contracting company that owns the rig, I'll list and explain all the job titles and what they involve.

Because many of the senior personnel on any rig have normally worked their way up through the ranks of the company, it makes sense to start at the entry-level positions and work our way up to the top roles. The role titles may change from rig to rig but the following list is a guide of the more common ones. This chapter will cover the deck crew and what they do.

Roustabout

The roustabouts are generally the least experienced and least skilled workers on the rig. It is the entry-level position for most people who start working on a rig who don't have any formal profession or trade. Generally a rigging background is an advantage because their main job is helping to move equipment around the decks.

Roustabouts have the most exposure to the weather than any other workers on the rig as they generally spend their entire 12-hour shift working on the open deck areas.

With the extreme remoteness of drilling locations also come extremes of climate. It's all too common to be working in conditions of extreme heat, extreme cold, extreme winds and/or extremely rough seas so you definitely need to have a tough skin and be physically fit.

According to Wikipedia: "An early 2010 survey by Careercast.com of the best and worst jobs — based on five criteria: environment, income, employment outlook, physical demands and stress — rated 'roustabout' as the worst job. Nonetheless, the anecdotal and subjective experience of an actual roustabout reveals the excitement of a challenging, adventurous job." If this is where you are starting off your career in the oil and gas industry then I guess it can only get better!

Given the limited amount of deck space on an offshore rig, only equipment that's needed for immediate operations are stored on the deck while other equipment is stored on nearby supply boats.

There's always a transfer of equipment going on between the supply boats and the rig and overhead heavy crane lifts are one of the most dangerous hazards that everyone has to watch out for while working offshore. Many deaths have occurred over the years from people mindlessly putting themselves at risk by being underneath a suspended load.

It is essential at all times that everyone walking around the decks watches out for overhead loads and stays outside the perimeter of where the load could fall should the suspension cables fail.

Because roustabouts don't require any previous skills they are generally sourced from the closest mainland base to where the rig is drilling offshore. With most countries these days requiring by law that international oil companies utilize a "Local Content" policy, the drilling company will usually source their unskilled laborers from the local national workforce.

Because of this, roustabouts are one of the most

transient work groups on an offshore rig. With most exploration wells, the roustabouts only stay on the rig for the duration of the drilling campaign (which could be just a few months or up to a few years) and if the rig then goes to another country for the next contract then the roustabouts will be laid off and new ones sourced at the next location.

Because of this fact, many roustabouts are employed through labor hire companies. In places where the producing fields are well established and long-term drilling is always being undertaken (for example many appraisal and development/production wells), the work continuity for roustabouts would be much more stable. Places like the Gulf of Mexico and the North Sea are two examples of this.

Like everyone who works outside the accommodation quarters, roustabouts have to adhere to the strict personal protective equipment (PPE) requirements. Basic mandatory requirements for PPE for workers on the deck include: hard hat, safety glasses, ear plugs, long sleeve shirt, long pants, (or coveralls), steel-capped work boots and impact gloves. This can vary from rig to rig but most rigs these days will have these as minimum mandatory PPE.

For rigs drilling in equatorial and sub-equatorial regions there is an added personal safety risk of people succumbing to heat illnesses because of these PPE requirements. Heat stroke can have fatal consequences and it's sometimes very hard to detect in workers as they will not want to be seen to be slacking off in their duties so will work through the signs of heat exhaustion until it escalates to heat stroke, by which time it can already have irreversible effects on the body.

There are many ways of managing this and supervisors need to be very aware of how the climatic conditions can be affecting the workers and use mitigating strategies to avoid the serious consequences of heat illnesses.

The roustabouts also oversee the helicopter operations. Whenever a chopper is due on deck they will be called into action to assist in ensuring the safety of incoming passengers disembarking, outgoing passengers boarding the aircraft, and the safe loading and unloading of bags and freight. They can also be required to refuel the chopper while it is on the deck, if necessary. The engines and rotors are rarely shut down and all operations are completed "on the fly". Accordingly, all tasks are undertaken with urgency and precision and follow a very strictly orchestrated routine.

While many roustabouts never aspire to be anything but a general laborer on the rig, many others use this start as a springboard to further their career in the industry and work their way up the ranks to a more senior drilling crew position.

Crane Operators

After the all-imposing derrick, the massive cranes are the most identifying features on any offshore drilling rig. All rigs have at least two large cranes but many will have three or even four.

Every piece of equipment offshore is heavy. Even the drill bits need to be lifted to the drill floor by cranes. Fifty tonne pedestal cranes are a heavy piece of equipment and deserve to be treated with respect. Being the workhorses of the rig means they are in constant use and therefore one of the highest areas for potential safety hazards. Regularly scheduled preventative maintenance is critical, as is having highly experienced people operating them.

Offshore crane operators have to be highly experienced to be able to handle the unusual work conditions. Unlike land-based crane operators who have a stationary worksite to operate on, offshore operators have to deal with the motion of the vessel they are setting down and picking up from.

There are six types of motion that a ship, or floating vessel, can experience and they are broken down into two categories, linear and rotational, and these each have three components to them.

Linear Motion

HEAVE – the linear vertical (up/down) motion.

SWAY – the linear lateral (side-to-side) motion, which is generated directly, either by the water and wind currents exerting forces against the hull, or by the vessels own propulsion.

SURGE – the linear longitudinal (front/back or bow/stern) motion imparted by the sea conditions.

Rotational Motion

PITCH – the up/down rotation of the vessel about its lateral axis (side-to-side). An offset or deviation from normal on this axis is referred to as "out of trim" and dynamically positioned rigs are constantly conducting trimming activities to keep the decks of the rig horizontally level.

ROLL – the tilting rotation of a vessel about its longitudinal (front-back) axis. An offset or deviation from normal on this axis is referred to as a list or heel. Heel refers to an offset that is intentional or expected, as caused by wind pressure caused by crew actions.

List normally refers to an unintentional or unexpected offset, as caused by flooding, shifting cargo, etc. With the ongoing loading and offloading of equipment and fluids from/to the supply boats this is always a motion that has to be compensated for. The rolling motion towards a steady state (or list) angle due to the ships own weight distribution is referred to as heel.

YAW – the turning rotation of a vessel about its vertical axis. An offset or deviation from normal on this axis is referred to as deviation or set.

It takes a very skillful operator to safely and successfully lay down sometimes very heavy and/or very large pieces of equipment in tight positions on a continually moving deck. Unpredictable wind and wave movements can make the task incredibly more difficult than the same load being handled on a land-based job.

Although the crane operators cabin is high off the deck they are still quite often working blind due to structures that obstruct their view of the deck where they are loading equipment. In these circumstances they rely totally on the verbal instructions over a radio or visual hand signals from a dogman on the deck that is in line-of-site of both the load and the crane operator.

All lifts are a team effort with precise and clear communication. Crane lifts in very strong winds – which are quite common offshore – can be very dangerous and really test the crane driver's and dogman's skills to the limit.

Although not legislated, it is generally expected that an offshore crane operator holds a Tower Crane (CT) High Risk License and has completed a certification course in operating an offshore crane. Prior experience as a dogman would almost always be the first step towards a job as a crane operator. Having experience working offshore doing other deck duties, such as a roustabout and dogman, would be essential prerequisites.

With efficient use of time being critical in all offshore operations it's essential that crane operators are highly skilled and able to maneuver loads quickly and precisely. There's little room for error and no time for repeated attempts.

All lifts are a well-coordinated collaboration between the crane operator, deck crew, third-party contractors who need their equipment moved and the DLC (drilling logistics coordinator). This "offline" process is ongoing 24 hours a day, 365 days of the year, to support the high-cost "online" process of drilling the well.

Much of the cost involved in drilling a well is related to the drilling operations and if these are delayed in any way by unprepared lifts of essential equipment then it starts costing the drilling company money. Being organized is key to all operations offshore. Time is money!

Deck Pusher

The typical term for a field supervisor or foreman in the oil and gas industry is a "pusher". The person in charge of supervising all the activities that take place on the decks of the rig is commonly referred to as the "deck pusher".

The deck pusher will have several years of experience working offshore and understand the logistics of moving equipment around the decks to tie in with the hectic pace of the drilling operations. He will coordinate the crews and permits required to plan the lifts and ensure they are all done safely and in a timely manner.

The deck pusher needs to have a thorough understanding of the rigs HS&E policies and also be very experienced in preparing and issuing work permits. Safety systems have to be very strictly adhered to while working offshore and the deck pusher must not only follow them himself but make sure all the deck workers he is supervising are following them too.

There will always be a day shift and night shift deck pusher on board the rig at all times as crane operations are ongoing 24 hours a day. The only boat-to-rig transfers that aren't generally carried out at night are fluid transfers via hoses. Things such as fuel and drilling fluids can contaminate the environment if they were to leak during the transfer so it's important to do these transfers during daylight hours so any leaks in the hoses or transfer system are detected immediately.

The deck pushers will meet daily with the drilling logistics coordinator (DLC) and the relevant third party personnel to plan each days lifts. Knowing where everything is placed on the decks at all times is of critical importance and the deck supervisor and DLC are in charge of managing this. Like all jobs on the rig, deck lifts rely on a detailed plan and team effort to get them done safely and in a timely manner.

****

This chapter has explained the general deck duties and the people responsible for getting them done. Chapter 5 will discuss the drilling crew and what jobs they are responsible for.

The drilling of an offshore oil and gas well is a very complicated and highly technical undertaking and you'll see in further chapters how the drill crew works in with third party contractors to get the job done. We're getting closer to the serious action now so stay tuned and keep reading!

CHAPTER 5

DRILLING CONTRACTOR – DRILL CREW

Now we're getting to the group of workers who are regarded as the "face" of the offshore drilling industry – the drilling crew. They are legendary not only because of the extremely physically demanding job that they do on the rig but also because of the shenanigans they are known to get up to when they aren't on the rig. If there's a drill crew stuck in town during a cyclone/hurricane evacuation then you can bet there's going to be a lot of alcohol drunk and crazy antics performed while they sit the storm out. These guys are not only known for their hard work but also their hard drinking.

While times are definitely changing, especially with the introduction of "zero tolerance" policies regarding alcohol and unsafe behavior, the drill crew still take pride in living up to their reputation of being tough both on the job and off it. They really are the lifeblood of the rig and the larger-than-life personalities of many of the people who work on the drillfloor make for interesting dynamics in the workplace.

Most of the senior drill crew members have worked offshore all their working life and have known nothing else but working and playing hard. If you're a quiet and sensitive kind of guy...then you don't belong in a drilling crew!

Despite this, there has been a subtle shift over recent years with the hardest workers also being the more health conscious on the rig and spending their off-tour time working out in the gym. With such a macho reputation to live up to it's important to also look the part!

Physical fitness is a very important aspect of a roughneck's job and staying in shape while on the rig is made more convenient now with all new rigbuilds seeing the importance of providing first class fitness facilities on the rigs.

It's also a common sight to see buckets of protein powders on the shelves in the galley as rig workers who are serious about their "gains" bring their muscle-building supplements to work with them. It's like a sub-culture within a culture that already demands a high level of physical fitness and resilience.

So what are the roles performed by the drilling crew? Lets take a look at them now, starting with the least experienced workers in the crew, the roughnecks.

Roughnecks

Roughnecks generally get their job through one of two channels:

1. Being promoted from a roustabout position after starting on the rig with no, or minimum, experience in the drilling industry.

2. Being employed with some experience in the drilling industry after working on land rigs. There is a lot more to offshore drilling than what there is on land rigs so even if a person has a lot of experience onshore they will still generally have to start at the beginning again when they go offshore.

Depending on the amount of previous relevant experience you had before joining the offshore drilling crew you could progress up the ranks quite quickly after getting a start as a roughneck. However, with each promotion up the ladder comes much higher responsibilities and many people prefer not to take on these more mentally demanding roles and are happy to stay a roughneck.

The main role of a roughneck is to assist in all areas of the drilling operations with the majority of their time being spent on the drillfloor. Everything on the drillfloor is very heavy and/or under very high pressures, making it one of the most dangerous places on the rig.

There is always something being run in the hole or being pulled out of the hole and the roughnecks are responsible for the physical work that is needed to get this done.

The driller operates the movement of the suspended drillstring but it's up to the roughnecks to maneuver the tubulars while connecting and disconnecting additional stands of drillpipe/casing/riser etc. Through all stages of the operation the roughnecks are involved in the manual handling of all drilling and specialist equipment.

Safety equipment is always evolving to minimize the dangers to drillfloor personnel through automation of many of the most hazardous tasks involved in the drilling process but stringent safety guidelines still have to be adhered to.

While machines like "iron roughnecks" are now commonly used to assist in pipe connections it's still paramount to always be aware of your surroundings. No amount of personal protective equipment (PPE) like gloves and hardhat, are going to save you from being crushed to death if you're standing in the wrong place when something "lets go" on the drillfloor.

Unpredictable occurrences like equipment malfunctions and downhole instability can happen at any time and have been the cause of many deaths in the industry over the years.

Many of the later generation rigs will have wireless communications between the driller and the roughnecks while performing operations on the drillfloor. Clear communication is essential to getting the job done quickly and safely and wireless headsets are becoming the norm now in many offshore operations.

With the driller operating from a room off to the side of the drillfloor it's necessary to communicate with the workers on the drillfloor via loudspeaker, with key personnel commonly also having wireless headsets so they are able to communicate directly with the driller.

Being the workhorses of the drilling crew means

the roughnecks don't generally make any of the decisions - they just have to follow the orders.

Having a high degree of commonsense and being able to communicate clearly are key qualities that a roughneck must have. Like roustabouts, the roughnecks are generally sourced from a relatively local workforce in the city closest to where the rig is operating. The more experienced they become, the more likely they are to get promoted to the next position in rank, which is generally the position of "derrickman".

Derrickman

The term "derrickman" was originally coined because one of their main tasks was to work up in the derrick of the rig and help maneuver the drill pipe in and out of the hole. This job was done from a platform high up in the derrick known as the "monkey board". This is still carried out today on many of the older rigs but the later generation rigs now have sophisticated pipe handling machines that automate a lot of these practices.

The derrickman's position is where the "brain" starts taking over from the "brawn". While still performing very physically demanding tasks, the derrickman now has a lot more responsibility than that of a roughneck and roustabout. As well as helping run and retrieve drillstring sections from the well bore either from aloft in the derrick or on the drillfloor, he is also responsible for monitoring and maintaining all the drilling fluid systems and equipment associated with the drilling of the well.

One of the most critical factors in the drilling of any offshore oil and gas well is the drilling fluid that is used to facilitate well bore stability and also to lift the drilled rock cuttings to the surface for analysis. The drilling fluid is commonly referred to as "mud" and it is a delicately balanced blend of chemicals and liquids with properties designed to optimize the drilling process. This mud is contained within a closed circuit throughout the drilling process and monitoring it is critical to the successful drilling of the well.

Not only do the properties of the mud have to be maintained but the volumes of the mud going in and coming out of the well have to be meticulously monitored to make sure the well isn't taking, or giving, any extra fluids.

Catastrophic events can follow either of those two situations if they aren't caught quickly enough. The Deepwater Horizon catastrophe occurred after an undetected increase in flow of the drilling fluid came out of the well. This was in direct response to hydrocarbons escaping uncontrollably from the bottom of the well, which eventually caused the devastating "blowout" that resulted in the deaths of 12 rig workers, the sinking of the rig, and the historical environmental disaster in the Gulf of Mexico in 2010.

The start location for the drilling mud is in the pit room, which is housed in a level below the drillfloor. While most onshore drilling rigs may only have 1 to 3 pits for mud, an offshore rig can have dozens of separate pits for storing mud for all contingencies, with each pit containing mud of differing properties.

These pits contain sensors for recording the volumes in each pit and these volumes are monitored by the derrickman. If he sees any discrepancies in the readings then he has to notify the driller immediately. (Many other people on the rig also monitor the pit levels, as you will read later in this book).

From the pit room, the mud is pumped via the mud pumps into the well, down the drillpipe, and then circulated back out to the surface. Once it reaches the surface it exits the flowline over a set of vibrating shaker screens that sieve the rock cuttings from the drilling fluid so the mud can then be re-used back down the hole.

The rock cuttings are centrifuged to recover as much of the mud as possible for reuse. The derrickman is responsible for the monitoring of this entire circuit of the drilling fluid.

Qualitative and quantitative analysis of the drilling mud is an important part of the derrickman's duties, as is the safe and efficient running of all the equipment involved in the process. Neglecting to see the warning signs of disequilibrium in the mud system can lead to catastrophic consequences…and very quickly.

You can start to understand now why some roughnecks just prefer to stay roughnecks! In reality though, it's not just the responsibility of the derrickman to keep an eye on the mud status but also that of many other people further up the line of command, as we will see in future chapters. And the next person further along that line of command is the Assistant Driller.

Assistant Driller

The assistant driller (AD) has worked his way up from either the roughneck or roustabout position and competently performed the role of a derrickman before getting promoted to officially start training as a driller.

He will work from the doghouse (the room the driller works from on the drillfloor) and will generally answer all phone and radio calls so the driller isn't disturbed from the serious business of drilling the well.

The drillfloor is usually a very busy place and the AD acts as an intermediary between the driller and more senior supervisors, and the derrickman and roughnecks. He will get trained in all aspects of well control but will not have the powers to act on any well control issues without being instructed to do so by the driller or more senior personnel.

Like the derrickman, the AD has to closely monitor the mud pits and alert the driller if there are any signs of gains or losses in the mud system.

All personnel from the assistant driller level and up have to have a current well control certification. The well control course is held over five days and the ticket is valid for 2 years. The entire 5-day course has to be repeated every time you need to be recertified.

Given the amount of experience needed to attain an assistant driller position, it is quite common for the AD's to be kept on for future drilling campaigns should the rig transit to a different country. The drilling company has had to invest a lot of time and money into training these members of the drill crew so they tend to hold onto them for as long as they can. Once they are deemed capable and competent, they can be promoted to the driller's chair at the next opportunity.

Driller

The driller probably has one of the least physically demanding jobs on the rig but one of the most mentally demanding jobs. While the drillers of days gone past would stand on their feet all day controlling "the brake" in a ramshackle corrugated tin shack, the operators on modern offshore rigs are now "cyber" drillers and control the whole drilling operations from a huge comfortable chair with joystick controls, in an air-conditioned room complete with coffee making facilities. Generally the only time they get out of the chair during their 12-hour shift is when they go for meal breaks.

The driller has the authority to take evasive action should he detect a serious well control issue that requires the well to be "shut in". The reaction time to detection and taking action may only be a matter of seconds so he has to have the knowledge and competency to be able to do whatever is needed to control the well. For this reason, it's essential that the driller isn't distracted with tasks that the assistant driller can otherwise take care of.

Along with the added responsibility comes the paperwork. The driller is responsible for filling in a mandatory IADC (International Association of Drilling Contractors) report every day which is a record of all matters pertaining to the drilling of the well for the two 12-hour shifts each day. The assistant driller and toolpusher will generally contribute to this requirement and between the three of them they will make sure all necessary activities are recorded after their shift has ended. These reports are submitted to a governing body and kept as a permanent record of the activities performed on the rig, regardless of the operations.

All drillers have worked their way up through the ranks so will generally have a minimum of several years of offshore experience. Because of the level of experience, and the cost in time and money it has taken to get them to this position, they will nearly always be permanent employees of the drilling contractor company. Once they have gained sufficient experience as a driller the next position up the chain of command is the drillfloor supervisor, or "toolpusher".

Toolpusher

The toolpusher is the supervisor that all the drilling crew reports to. He oversees all aspects of the drilling operations and is the intermediary between the drilling crew and the rig manager (OIM) and the company man. There will generally be a dayshift and nightshift toolpusher, and sometimes an additional senior toolpusher in large operations.

The toolpusher will spend most of his time in the doghouse assisting the driller during critical times of the drilling operations and when things are quiet on the drillfloor you'll find him in an office in the accommodation quarters filling in paperwork and replying to emails. As with any job, the higher up the food chain you get, the less physical work you have to do but get overloaded with additional responsibilities and administrative tasks.

The computer age has well and truly hit the offshore drilling industry, with computer systems controlling all aspects of the drilling operations on modern offshore rigs.

There are dozens of sensors placed all over the rig, monitoring all aspects of the drilling operations. Voluminous amounts of data are collected from sensors that record drilling parameters (such as torque on pipe, overpull, hook height, rate of penetration, drillpipe revs, weight on bit, etc.), circulating mud properties (pump rate, mud temperature in and out of the hole, mud weight, circulating density, static density, gas content, etc.) and also sensors that detect gas and other contaminants in the air that have been circulated out of the well.

Computer systems all over the rig monitor the outputs of these sensors and alarms will warn of changes in parameters or outputs that can be signals of unfavorable conditions. Daily reports have to be produced by all departments and the data collected all over the rig is used to monitor the progress of the drilling operations. The toolpusher has to make sure all of these monitoring capabilities are operative and manage the ongoing maintenance and operation of all rig data collection equipment.

Most of the computerized systems only record data and warn of any possible dangers, with human intervention needed to take evasive action. However, there are also automated systems that initiate evasive action should dangerous situations be detected.

One of these is the Deadman Auto Shear (DMAS) that can activate the closing of the shear rams on the blowout preventer (BOP) automatically, based on too high pressure or excessive flow. While the toolpusher has to make sure all these systems are operative, the responsibility of maintaining and operating these systems falls under the workscope of the Subsea department, which will get explained in a later chapter.

There's only one more person on the rig that holds a higher position than the toolpusher within the drilling crew and that is the OIM.

Offshore Installation Manager (OIM)

The OIM is responsible for all the personnel on the rig and for the safe drilling of the well. The buck stops with him in regards to any drilling, machinery or personnel issues on the rig that fall under the drilling contractors responsibilities.

He is the big daddy of report writing and will spend most of his day working in his office fulfilling mandatory reporting requirements, answering phone calls, replying to emails and attending meetings.

Together with the company man, the OIM must be consulted before any procedural changes can be made on the rig.

The OIM will generally have worked his way up the ranks from a roustabout or roughneck position and will have decades of experience in the offshore drilling industry. In emergency situations the fate of the rig, and all people working on it, will ultimately be his responsibility. There is only ever one OIM on board – one of the few positions on a rig that doesn't have a dayshift *and* nightshift representative.

****

In Chapter 6 I'll explain roles that are performed by the drilling contractor specialists in the subsea department. These crews are working in the background throughout the drilling operations and are critical to ensuring a safe work environment beneath the water.

CHAPTER 6

DRILLING CONTRACTOR – SUBSEA CREW

In this chapter we'll explore the crew that works with all the equipment and operations that are performed between the drillfloor and the seabed. The "SUBSEA" crew is employed by the drilling contractor and is an integral part of the offshore operations.

Subsea Operations

The subsea crew is responsible for implementing

and maintaining the structures, tools and equipment used in the underwater components of offshore oil and gas drilling and production operations.

The underwater environment presents unique challenges to subsea engineers, particularly deepwater operations where temperature, pressure and corrosion test the durability of submerged equipment and tools. Most subsea engineering operations depend on automation and remote procedures to construct, maintain and repair components beneath the surface of the water.

To understand what tasks the subsea team are required to undertake we first need to explore the key structures between the seabed and the drillfloor that connect the drilling unit to the well bore.

Up until now we've only been looking at the elements of offshore drilling that lie above the water line but there's also a lot of technology hiding beneath the surface of the water. Starting from the seabed and working our way up to the drillfloor we'll look at the subsea components that help us bring drill cuttings and potentially trapped hydrocarbons safely to surface.

With the deepest-water offshore well ever to be drilled lying in 3,400 m (11,155 ft) of water, it's easy to see why a team of specialists needs to be employed to oversee the operations that happen beneath the waves.

Wellhead

The subsea wellhead system is a pressure-containing vessel that provides a means to hang off and seal off casing used in drilling the well. The wellhead also provides a profile to latch the subsea blowout preventer (BOP) stack and drilling riser back to the floating drilling rig. In this way, access to the wellbore is secure in a pressure-controlled environment. The subsea wellhead system is located on the ocean floor, and must be installed remotely with running tools and drillpipe.

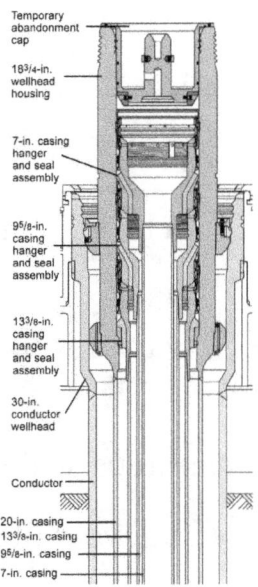

Temporary abandonment cap

18³/4-in. wellhead housing

7-in. casing hanger and seal assembly

9⁵/8-in. casing hanger and seal assembly

13³/8-in. casing hanger and seal assembly

30-in. conductor wellhead

Conductor

20-in. casing
13³/8-in. casing
9⁵/8-in. casing
7-in. casing

The subsea wellhead inside diameter (ID) is designed with a landing shoulder located in the bottom section of the wellhead body. Subsequent casing hangers land on the previous casing hanger installed. Casing is suspended from each casing-hanger top, and accumulates on the primary landing shoulder located in the ID of the subsea wellhead. Each casing hanger is sealed off against the ID of the wellhead housing and the outside diameter (OD) of the hanger itself with a seal assembly that incorporates a true metal-to-metal seal. This seal assembly provides a pressure barrier between casing strings, which are suspended in the 18¾-in. wellhead.

A standard subsea wellhead system will typically consist of the following:

- Drilling guide base.

- Low-pressure housing.

- High-pressure wellhead housing (typically 18¾ in.).

- Casing hangers (various sizes, depending on casing program).

- Metal-to-metal annulus sealing assembly.

- Bore protectors and wear bushings.

- Running and test tools.

The drilling guide base provides a means for guiding and aligning the BOP onto the wellhead. Guide wires from the rig are attached to the guideposts of the base, and the wires are run subsea with the base to provide guidance from the rig down to the wellhead system.

Blowout Preventer (BOP)

There are two means to prevent an escape of high-pressure fluids or gases from the well when drilling for oil and gas.

The primary means is hydrostatic pressure with weighted up drilling mud and the secondary means is the blowout preventer. The BOP is literally the last line of defense in preventing a catastrophic event on the rig.

The BOP is an arrangement of valves, rams preventers, annular preventers, connectors and control system that can be controlled from the surface to "shut-in" the well in the event of an impending blowout.

In addition to controlling the downhole pressure and the flow of oil and gas, blowout preventers are intended to prevent tubing, tools and drilling fluid from being blown out of the wellbore when a blowout threatens. Blowout preventers are critical to the safety of crew, rig and environment, and to the monitoring and maintenance of well integrity.

With the wellhead just above the mudline on the sea floor, there are four primary ways by which a BOP can be controlled. The possible means are:

- Electrical Control Signal: sent from the surface through a control cable;

- Acoustical Control Signal: sent from the surface based on a modulated/encoded pulse of sound transmitted by an underwater transducer;

- ROV Intervention: remotely operated vehicles (ROVs) mechanically control valves and provide hydraulic pressure to the stack (via "hot stab" panels);

- Deadman Switch / Auto Shear: fail-safe activation of selected BOPs during an emergency, and if the control, power and hydraulic lines have been severed.

Two control pods are provided on the BOP for redundancy. Electrical signal control of the pods is primary. Acoustical, ROV intervention and deadman controls are secondary.

An emergency disconnect system, or EDS, disconnects the rig from the well in case of an emergency. The EDS is also intended to automatically trigger the deadman switch, which closes the BOP, kill and choke valves. The EDS may be a subsystem of the BOP stack's control pods or separate.

Pumps on the rig normally deliver pressure to the blowout preventer stack through hydraulic lines. Hydraulic accumulators on the BOP stack enable closure of blowout preventers even if the BOP stack is disconnected from the rig. It is also possible to trigger the closing of BOPs automatically based on too high pressure or excessive flow.

The subsea team is responsible for all maintenance and testing of the BOP and it's ancillary equipment. Function tests are carried out frequently throughout the drilling program, especially prior to running "the stack" from surface, and also prior to drilling through expected reservoir formations.

The drilling crew and subsea team run coordinated tests from both the drillfloor and the backup system's control panel within the accommodation unit. Every rig must have a BOP control panel at the driller's station as well as one in a safe location away from the drill floor.

The members of a subsea team are generally recruited with an electrical or mechanical trade base or engineering degree and they then go through extensive training programs to familiarize themselves with the subsea operations. Because of the skills required to be able to competently do their job these crew members don't start working offshore as an unskilled laborer like many of the drilling crew members generally do. Subsea operations are a highly specialized field and as such, highly specialized teams are required to perform the tasks involved.

It is also one of the most highly regulated areas in the offshore drilling industry due to the fact that failures in the system can result in catastrophic events, such as the Deepwater Horizon disaster. Being the last line of defense in the event of a blowout, it is critical that all the subsea equipment can be reliably called upon to shut the well in during a well control emergency situation.

Because the BOP is such a critical part of the process safety systems offshore, since the Macondo blowout there have been strict regulatory requirements imposed on the industry to ensure the operators have clear programs in place to identify potential hazards when they drill, clear protocol for addressing those hazards, and strong procedures and risk-reduction strategies for all phases of activity, from well design and construction to operation, maintenance, and decommissioning.

Adhering to these regulations requires certification of all subsea equipment from an independent third party regarding the condition, operability, and suitability of the BOP equipment for the intended use and the operator must have all well casing designs and cementing program/procedures certified by a professional engineer, verifying the casing design is appropriate for the purpose for which it is intended under expected wellbore conditions.

Third-party verification and inspection organizations (such as Subsea Solutions http://www.subseasolutions.net) work with subsea equipment, specifically BOP and regulatory compliance audits, well-control and drilling equipment inspections, to ensure the highest levels of integrity within the subsea well control system prior to it being deployed.

Adjoining the top of the BOP and connecting with the bottom of the marine riser is the lower marine riser package.

Lower Marine Riser Package (LMRP)

The LMRP is the upper section of a two-section subsea BOP stack consisting of the hydraulic connector, annular BOP, ball/flex joint, riser adapter, jumper hoses for the choke, kill and auxiliary lines and subsea control modules. The LMRP interfaces with the BOP stack.

Blowout preventers must have completely redundant control systems on the BOP. These control systems are called pods and are designated Blue Pod and Yellow Pod in all systems, no matter which manufacturer. They can be found on the lower marine riser package and are extensively function tested prior to the deployment of the BOP.

There can be as many as six emergency systems in a BOP to operate critical functions in the case of the loss of the primary control system:

1. **Emergency Disconnect Sequence** (EDS) – In a case where a dynamically positioned rig has lost station-keeping ability, the EDS is a one button system that allows the wellbore to be secured by closing the shear rams. The hydraulic functions to the lower BOP are then vented and the LMRP is separated from the lower BOP by unlatching the connector. An over-pull is preset on the riser tensioners and the LMRP lifts from the lower BOP. A riser recoil system prevents a sling shot effect. After the EDS button is activated, the sequence takes about 55 seconds maximum.

2. **Acoustic systems** – A limited number of emergency functions (typically shear rams and LMRP connector) can be operated from the rig using a hydrophone transmitting to transducers on the BOP. It is uncertain if these systems will work in a well control situation where considerable noise is

generated from flow in the wellbore.

3. **Remote operated vehicles (ROVs)** have pumps which can operate functions through a 'hot stab' plugged into a dedicated receptacle in panel. The limitation of an ROV is the time to deploy from the rig to the seabed and the limited flow rate of their pumps.

4. **Dead man systems** will close the shear rams in the event all hydraulic and electric control is lost on the BOP. This would typically only happen if the riser string parted. In deepwater if the riser is lost, then the hydrostatic pressure of the drilling mud, which is needed to contain wellbore pressure, would be reduced as it is replaced by seawater. Closing the shear rams secures the well.

5. **Automatic Disconnect System** (ADS) closes the shear rams when the lower flex joint reaches a preset angle.

6. **Autoshear** closes the shear rams in the event the LMRP is unintentionally disconnected.

The BOP and LMRP are run subsea using the "marine drilling riser" after the surface well has been drilled and a wellhead has been landed and cemented in the seabed.

Marine Drilling Riser and Marine Riser Tensioner

A marine drilling riser is a conduit that provides a temporary extension of the subsea oil well to the drilling rig. The "riser" has a large diameter, low pressure main tube with external auxiliary lines that include high pressure choke and kill lines for circulating fluids to the subsea blowout preventer (BOP), and usually power and control lines for the BOP.

When used in water depths greater than about 20 meters, the marine drilling riser has to be tensioned to maintain stability.

A marine riser tensioner located on the drilling platform provides a near constant tension force adequate to maintain the stability of the riser in the offshore environment. The level of tension required is related to the weight of the riser equipment, the buoyancy of the riser, the forces from waves and currents, the weight of the internal fluids, and an adequate allowance for equipment failures.

The marine riser is kept in tension with large pistons operated with an air/oil system at pressures up to 3,000 psi. The riser may be connected via a tensioning ring to wire rope, which is reeved over sheaves on the pistons, or the pistons may be connected directly to the riser tensioner ring.

Once the BOP stack has been successfully run to the seabed with the marine riser and latched onto the wellhead, it will undergo another series of function tests to determine its operability under water depth conditions.

The density of water can cause problems that can increase dramatically with depth. The hydrostatic pressure at surface is 14.6 psi (pounds per square inch) but this increases by this amount for every 10 metres of water depth. For a deepwater well that has the wellhead on the seabed in 2,000 m of water you would expect to find the hydrostatic pressure acting on the BOP to be around 3,000 psi.

When you also consider the water temperature to be close to 0° Celsius then you can imagine the type of hostile environment these safety-critical components have to function under. Making equipment that can operate under these conditions is the job of the manufacturers subsea engineers – making sure they work and keeping them well maintained is the responsibility of the subsea engineers onboard the rig.

Troubleshooting BOP issues is generally collaboration between specialist subsea engineers onshore and the subsea maintenance crew involved in the offshore operations. If subsea function tests fail then the entire BOP stack and riser string has to be pulled up to surface so physical examination of the unit can take place.

This is a very time-consuming and costly exercise so making sure everything is functioning 100% before running it down to the seabed is imperative. As anyone who has ever worked offshore knows, it's all-too-common for BOP's to fail function tests and this is why such strict regulatory conditions have been placed on the subsea components used for the drilling of offshore wells, especially in deepwater and ultra-deepwater wells.

Once the BOP has been successfully tested it's time to drill ahead!

CHAPTER 7

Offshore Oil and Gas

PEOPLE

DRILLING CONTRACTOR – MARINE AND MAINTENANCE CREWS

The drilling contractor not only operates, and is in charge of, the drilling operations in offshore drilling operations but they are also the owners and operators of the vessel that is being used to drill the wells.

Particularly in the case of drillships, they need to employ a full compliment of marine crew who run the vessel while the drilling is taking place.

Like all sea-going vessels there are a number of ranks, which are recognized positions on most offshore oil and gas drilling rigs. Due to the nature of jack-up rigs being non-floating rigs, they will have minimal marine crew as they get towed from one location to the next. The more autonomous a drilling unit becomes (i.e. the less dependent it is on external forces for propulsion and stability), the greater the need for a full compliment of marine crew.

The marine crew can generally be divided into four main categories: the bridge, the deck, the engineering department and the steward's department. Depending on the type of rig, you may find all or only some of the following personnel working on a rig. Given that their duties are solely to do with the running of the vessel and not the drilling of the well, they nearly always will come from a marine industry background rather than a drilling industry background.

Captain/Master

The "captain" or "master" is the vessels highest responsible officer, acting on behalf of the rig's owner. The captain is legally responsible for the day-to-day affairs of the rig as they are in command. It is their responsibility to ensure that all the departments under them perform legally to the requirements of the ship's owner.

The captain/master will have his own cabin on the rig that contains his bedroom and office. He'll generally work out of here with his job involving mainly administrative duties, which keeps him tied to a computer for much of the day. Like all senior roles on a rig, he gets heavily bogged down with daily reports, on-site meetings, phone calls, conference calls and emails.

Chief Engineer

The chief engineer is responsible for keeping the ship and the machinery running. Today's mobile offshore drilling units are complex vessels that combine a lot of technology within a small space.

This includes not only the engine and the propulsion system, but also, for example, the electrical power supply, devices for loading and discharging, garbage incineration and fresh water generators. The chief engineer is responsible for all operations and maintenance that have to do with all machinery and equipment throughout the vessel.

Dynamic Positioning Operator (DPO)

The bridge on any rig is filled with sophisticated navigational equipment, with 7th Generation drillships having the most advanced systems in the offshore industry.

To maintain their position, drillships and semi-submersible rigs may utilize their anchors or use the ship's computer-controlled system on board to run off their dynamic positioning.

DPO's are the people who are in charge of controlling the Dynamic Positioning System. The purpose of this system is to automatically maintain a vessels position and heading by using its own propellers and thrusters. This is a very complex task because DP systems need to combine position reference sensors, wind sensors and motion sensors to calculate the impact of environmental forces that affect the vessels position.

DP systems are vital for safely carrying out operations in water too deep for stabilization using anchors or jack-up legs.

The system was created in the 1960's to meet the demands of the oil & gas industry and allowed offshore drilling units to operate in deeper waters than was previously done. This paved the way for the discovery of new fields and gave birth to the deepwater and ultra-deepwater drilling industry. Dynamic positioning has many advantages, such as excellent maneuverability, no additional vessels required to work with anchors, able to operate at any water depth, quick set-up and not limited by an obstructed seabed.

There can be very serious consequences resulting from the loss of position of the floating rig, the main ones being:

- The rig could disconnect from the subsea wellhead, BOP or marine riser, which could cause uncontrolled oil spills and possibly create serious environmental problems.

- If there are crane operations taking place with a supply boat alongside the rig then the possibility of a collision with the two vessels is very high and could result in serious damage to either vessel as well as the risk of injuries or even deaths of personnel working on them.

- If there are divers working beneath the rig they are completely dependent on the vessel while working underwater so if the rig was to lose position it could have fatal consequences for the personnel working under it.

The computer program contains a mathematical model of the vessel that includes information pertaining to the wind and current drag of the vessel and location of the thrusters. This knowledge, combined with the sensor information, allows the computer to calculate the required steering angle and thruster output for each thruster.

Dynamic positioning may either be absolute in that the position is locked to a fixed point over the bottom, or relative to a moving object like another ship or an underwater vehicle. One may also position the ship at a favorable angle towards wind, waves and current, called "weathervaning".

Dynamic Positioned vessels are described as being Class 1, Class 2 or Class 3.

Equipment Class 1 (DPS-1) has no redundancy - Loss of position may occur in the event of a single fault.

Equipment Class 2 (DPS-2) has redundancy so that no single fault in an active system will cause the system to fail. Loss of position should not occur from a single fault of an active component or system such as generators, thruster, switchboards, remote controlled valves etc., but may occur after failure of a static component such as cables, pipes, manual valves etc.

Equipment Class 3 (DPS-3) also has to withstand fire or flood in any one compartment without the system failing. Loss of position should not occur from any single failure including a completely burnt fire sub division or flooded watertight compartment.

Unlike the rest of the workplaces on the rig, the bridge is a spacious, quiet, clean office with the best views in the "building". It looks more like the flight deck on the "Starship Enterprise" than a bridge on a drilling rig. Everyone appears to talk in hushed tones and keeps to themself – well that's what it seems like after working out on the deck where everyone is screaming at each other to be heard over the noise of the rig.

Because the loss of position may cause financial, environment, health and safety risks it is essential that only highly qualified personnel can control DP systems. There will generally be two DPO's on-tour at any one time (one of which will be a senior DPO), so there is always someone "on watch" while the other has meal/bathroom breaks.

The DPO has to be competent to use the DP systems in manual and automatic modes without supervision. There is a special training scheme for achieving safety standards for DPO's in the offshore oil & gas industry. This scheme defines the basic stages of professional training for future DPO's with three main blocks of instruction – practice on board a DP vessel, theoretical sessions and simulator training at special training centres. The DP "UNLIMITED" certificate will be issued after the successful completion of all phases of the training scheme. This certificate confirms the competence of an operator and allows him to work on DP systems without supervision.

The DPO will generally also assist in the co-ordination of all deck, crane, gangway, and helicopter and supply boat operations, as well as safety operations that are carried out from the bridge, such as emergency shut-in procedures and safety drills.

Radio Operators

The radio operator generally works out of the bridge, alongside the DPOs, and they will generally be the first point of call in any emergency situation. Their main role is to provide reliable communications between the vessel and the shore, other vessels and helicopter traffic.

The radio operator works under the supervision of the Captain/Master, and reports directly to them. They establish and maintain the 'flight watch' during all helicopter operations, and record all details of landings and departures of aircraft.

They also keep detailed records of all the persons on board (POB) and what cabin everyone is sleeping in. On fly-out day all departing crew have to report their bag and body weights to the radio operator so they can provide an accurate manifest for the outgoing chopper during crew change operations.

The maintenance of all radio equipment and emergency power sources, stock keeping of radio spares, etc. are all an integral part of the Radio Operators role. In addition, the Radio Operator assists the Master with general clerical work such as vessel documentation, daily reports, timesheets and meeting minutes, so they must be computer literate. The Radio Operator also takes on a support role in emergency situations, in both a communication role and as part of the vessels emergency command and control team. This involves assisting with POB reconciliation; log keeping, emergency radio communications, etc.

Because of the prerequisite marine certifications that are required for radio operators, it is quite common for ex-navy personnel to be working in these positions. There is always a nightshift and dayshift radio operator on the rig and when they need to have a food or comfort break throughout their shift they will get the medic or someone else to cover for them so the phone is always manned for emergency calls.

With the radio operators job being so critical for communications to and from the rig it is no wonder satellite communications equipment is an integral part of offshore operations these days. Not only is this equipment necessary for the receipt of weather warnings, transmission of position reports, priority traffic, distress messages and just keeping in touch with family and friends, but also the dissemination of well data to and from the rig.

Enormous amounts of drilling data is sent in real time to the head office onshore so the specialists and project managers working in the head offices of the energy companies can view the data via a live feed from the rig. All major decisions made on the rig are nearly always in consultation with project managers onshore, so a reliable communication network is essential for timely decisions to be made. Remember...time is money...BIG money!

With the improvements in offshore communication technologies comes an improved work-life balance for offshore workers. Now, offshore personnel are able to use wifi, talk to family and friends on the phone and watch television during their off-tour time. If the Internet goes down you'll soon hear about it! And it quite often does go down; severe weather plays havoc with the satellite hardware and also the satellite TV reception...just like your Foxtel does during a severe storm when you're at home.

Ballast Control Operators (BCO)

Semi-submersible rigs (and to a lesser degree Jack-ups) also need operators to control the buoyancy of the rig from the ballasted, watertight pontoons located below the ocean surface. With its hull structure submerged at a deep draft, the semi-submersible is less affected by wave loadings than a normal ship but with a small water-plane area, the semisub is sensitive to load changes, and therefore must be carefully trimmed to maintain stability.

Semisubs are able to transform from a deep to a shallow draft by deballasting (removing ballast water from the hull), thereby becoming surface vessels. Usually they are moved from location to location in this configuration.

Jack-up rigs also need a certain degree of ballast control as they have large ballast tanks built into the structure. When the rig is jacked down, its hull floats on the surface of the water like a ship. The ballast tanks can be flooded with water or pumped free of water to control its buoyancy.

The BCO is the designated person in charge to maintain stability of the rig and evaluate the possible effect of load combinations while on station and when undertow. He has to maintain the rig at the required operating draft and keep it upright and on an even keel unless otherwise requested by the OIM / Barge Engineer/Master. Other responsibilities of the BCO include:

- Evaluate the possible effects of load combinations on the rig during unloading and back loading of supply vessels when the rig is under tow or on station

- Assist the Barge Engineer during anchor handling, shifting and moving operations.

- Supervises all major changes in deck load distribution, as well as shifting, loading and off-loading of fluids to and from the rigs tanks.

- Assist the Barge Engineer in maintaining rig drawings and other documents related to the structure and equipment of the rig.

- Trim the rig whenever required by ballasting or deballasting.

Complete standard control room logbook, daily ballast report and official logbook.

Keeping the rig trimmed so the deck is perfectly horizontal is critical to drilling operations as the laser sensors on the mud pits will read incorrect volumes if the pits are not level. This effect could mask potential dangers of the well either "kicking" or "taking a drink", in which the well either has too much mud coming back out of the hole or it loses mud into the formation – both of which can have severe consequences.

It is critical that the BCO makes a general announcement over the rigs PA system to alert everyone on the rig if they are "trimming the rig", so people know to account for the corrections should it affect the mud pit volumes or any drilling parameters.

An even more critical reason to maintain control of the ballast of the rig is to prevent any catastrophic "listing" events. Semisubs have been known to suffer from such severe listing from errors in ballast control they have literally sunk into the sea. No one ever wants to see this happening while they're on the rig!

Mechanics and Electricians

All rigs have a chief mechanic and chief electrician who lead the electrical and mechanical teams onboard. With offshore rigs being heavily mechanized and automated, both teams are kept very busy. It's not uncommon to find rig mechanics branching into other fields after they get experience on a rig and find career progressions with specialist third party contracting companies (which will be the topic of the next article).

Medics

There will nearly always be a dayshift and nightshift medic onboard and generally at least one of them will be a fully qualified medical doctor. It's quite common for rig medics to have a military background as they are well trained in emergency medical care and used to working in harsh environments and away from home for long periods of time.

While the drilling contracting company is responsible for employing them they will nearly always be sourced from a third-party agency and not directly employed through the drilling company.

The medics work out of the rig "hospital" which is equipped to handle comprehensive first aid care and emergency medical procedures when needed.

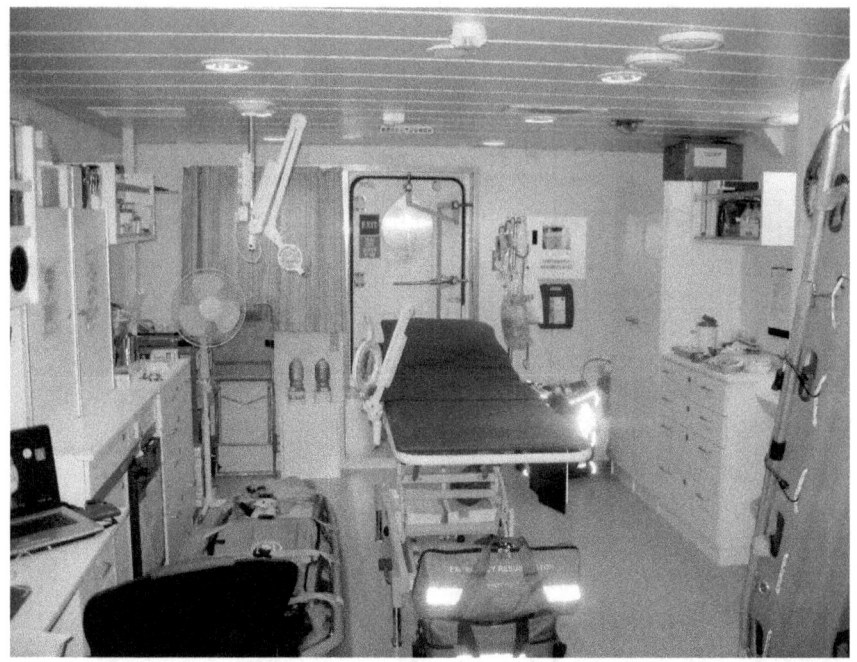

Rig Safety and Training Coordinator (RSTC)

The RSTC is the person responsible for ensuring that all tasks on the rig are completed in accordance with company and regulatory requirements by using approved procedures and permits.

Safety reps normally come from quite varied backgrounds, with many having worked other roles within the drilling industry or sometimes even come from a military background. A background as a rig worker is most advantageous because they would then have a competent knowledge of the tasks performed on the rig as well as all the equipment being used. This knowledge would make investigations and report writing of incidents a lot easier.

The main duties include:

- Assist and coach rig-based personnel in the implementation and understanding of the Company Management System

- Act as installation subject matter expert (SME) for the Company HSE management system and training requirements

- Provide a communication link between QHSE & Training organizations and the offshore workforce

- Assist in the implementation of the company goals and objectives through liaison with QHSE Management

Materials Coordinator (Storeman)

Storeman is an entry-level position that requires no prior offshore experience, although working in a similar role at onshore operations would be an advantage.

They are responsible for the storage and

distribution of consumable items on the rig, ranging from mechanical and electrical equipment to personal protective equipment. They will also oversee the offloading and storage of chemicals that are needed for drilling operations, which get stored in either the store area or the sack room.

****

As you can imagine, there would be many other roles involved in offshore drilling operations that I wouldn't have covered in this chapter but I have provided an overview of the main ones that all rigs generally have. In the next chapter we'll start to look at the many third-party contractor roles that are performed on the rig.

CHAPTER 8

THIRD-PARTY SERVICE PROVIDERS FORMATION EVALUATION AND WELL MONITORING

The third party service providers make up about half of the workforce on an offshore rig. With so many hi-tech and specialized operations being performed at all stages of the drilling operations it's imperative that experts in their field perform these tasks.

The first group of people we'll look at are the "mudloggers" who's job it is to monitor the drilling operations from the time they "spud" the well to the time they secure the well after drilling and testing has been completed.

Mudloggers

"Mudlogger" is the generic term used to describe the field specialists who monitor the well and also collect samples for the geologist. The career progression for a "mudlogger" is to generally start as a sample catcher while they learn about the drilling operations, then progress to a mudlogger and with further experience, become a data engineer.

Sample Catchers

Dedicated sample catchers aren't always part of the team but they often get "thrown in" as a complimentary part of the mudlogging services. They don't need to have any prior experience in working offshore or as a mudlogger, so it's a very good entry level job and is generally the starting position for graduate geologists who wish to work offshore. Although you don't need to be a geologist to be a sample catcher, most of them will be and will go on to get trained as a mudlogger.

Sample catching is without a doubt the least glamorous and lowest paid of all jobs on the rig…but you have to start somewhere! The role of a sample catcher is to provide the most basic geological data acquisition on the rig and to assist with all general activities when possible. The main duties of the sample catcher are:

- Ensuring that representative geologic samples are caught throughout the drilling or reaming phases of the well program. This is done by collecting cuttings (drilled rock) samples, from the proper "lagged" (explained below) depths and at the proper intervals as required for evaluation. These samples are collected off the shale shakers, screened and washed, divided into correct portions, and packed into sets for the Client, partners and government agencies. They may also have to assist in core recovery and packaging as required.

- Preparing a clean "cuttings" sample on a sample tray for the wellsite geologist and mudlogger, who will then examine it under the microscope and describe the lithology of the drilled formation.

- Assisting mudloggers and data engineers to perform regular and frequent calibration checks of instruments, perform normal routine maintenance of sensors and other equipment and also assist logging crew with rig-up/rig-down procedures.

The sample catcher reports directly to the mudlogging crew who will ensure his duties are performed correctly. This may include on-the-job training as required. They work out of the mudlogging unit, which is always close to the shale shakers and these are generally one or two levels below the drillfloor.

The shale shakers are vibrating screens that separate the drilling fluid from the drilled rock cuttings. The "shaker house" is a very noisy place and double hearing protection must always be worn. There will be multiple shakers to accommodate the large volume of cuttings that can be produced when the drilling rate of penetration is high (i.e. they are drilling fast!). It's a very "dirty" job and multiple layers of personal protective equipment need to be worn to prevent skin contact with the drilling mud, which can cause serious skin inflammation.

Mudloggers and Data Engineers (DE)

Mudloggers and data engineers are responsible for gathering, processing and monitoring information pertaining to drilling operations. They don't only collect data using specialist data acquisition techniques – they also collect oil samples and detect gases using state-of-the-art equipment.

The information amassed by these guys is analyzed, logged and then communicated to the team that is responsible for the physical drilling of the well. Without the help of the mudlogger, the drilling operations would be less efficient, less cost-effective and much more dangerous. The mudlogger is vital for preventing hazardous situations, such as well blowouts.

They also provide vital assistance to wellsite geologists and write detailed reports based on the data that is collected. Being an entry-level position, employees will be given a mixture of 'on-the-job' training and expert in-house training courses, which cover different aspects of drilling operations. A major part of the training will focus on the use of specialist computer software.

Typically, you will need a degree in geology to start a career as a mudlogger. However, candidates with degrees in physics, geochemistry, chemistry, environmental geoscience, maths or engineering may also be accepted.

Along with the sample catchers and data engineers, the mudloggers work out of the muddlogging unit, which is a pressurized sea container-type of office, which is positioned close to the drillfloor and shaker house.

The unit will have an air-lock compartment when you first enter it so as to maintain the positive pressure within the unit whenever somebody leaves or enters the unit.

This is the main control room for monitoring the drilling operations and is full of sophisticated and delicate equipment and computer systems. Positive pressure needs to be maintained to ensure the air pressure inside the container is higher than that of the outside area to prevent contamination of sensitive monitoring equipment – and also to ensure the safety of the crew working inside the unit should the outside air become contaminated through uncontrolled releases of hydrocarbons from the well.

One of the most important tasks of the mudlogger is to oversee the collection of not only geological samples but also mud and gas samples from the well during drilling operations. To be able to do this accurately they have to know the exact "lag time" (or "bottoms-up time") that it will take for the drilled cuttings or mud and gas to arrive at the surface after being drilled and circulated up the outside of the drillhole (annulus) while suspended in the drilling mud. The lag time may be a few minutes in a shallow hole or as much as several hours in deep wells with low mud flow rates. To be able to work this time out accurately there are many factors that have to be taken into consideration. The lag time depends on:

- the annular volume fluid

- flow rate, which in turn require knowledge of:

- dimensions (internal diameter (ID) and outside diameter (OD)) of surface equipment, drill string tubulars and casing and riser.

- mud pump output per stroke, pumping rate and efficiency.

While the computer's software will work this out automatically, the calculated value may be incorrect however, if the operator has entered erroneous or incomplete values for the pipe or hole dimensions, or if the hole is badly washed out. This has to be monitored very carefully to avoid catching mud, gas and cuttings samples at incorrect depths.

Sensors

The mudloggers and DEs monitor the drilling operations via a series of sensors that are placed at various locations around the drillfloor, pit room and shaker house.

The main drilling and mud parameters that are

recorded are: hook movement, weight on hook, standpipe pressure, wellhead pressure, rotary torque, pump strokes, RPM, mud pit levels, mud density, mud temperature, mud resistivity and mud flow.

These parameters are monitored in real time and any deviances from the expected normal values must be immediately reported to the driller. The DE will view and monitor all the drilling parameters on a screen as shown below.

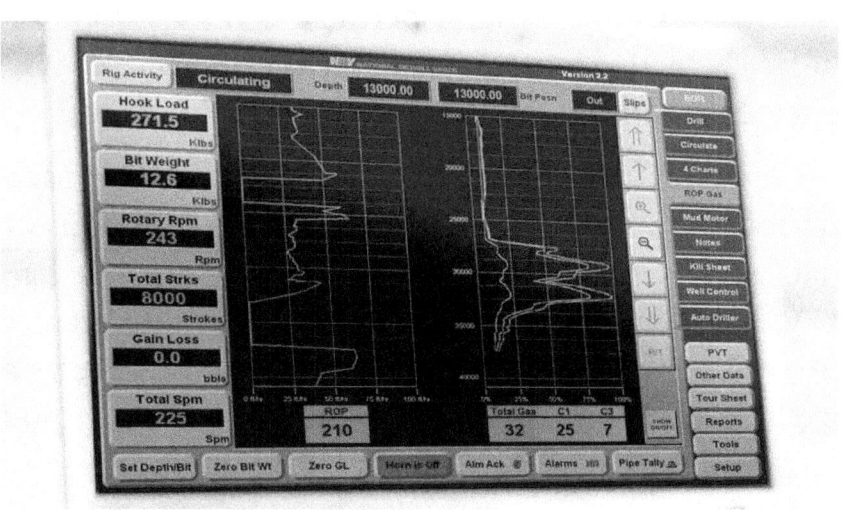

The five most important monitoring tasks that the mudlogger and DE must watch out for are:

- Rate of penetration increase, which could

indicate they have drilled into a reservoir formation

- Mud pit volume gain or loss, which could indicate the well is taking a kick, or losing fluid into the formation

- Mud flow rate change

- Mud density variation

- Indication of oil or gas.

The mudlogging unit is a very confined workplace and there may be up to several people working in there at any one time, especially if it's a "combo" unit, which houses the mudloggers, MWD engineers and possibly also the directional driller.

Generally the same service provider company performs all of these roles so it is quite common for data engineers to progress into a role as an LWD/MWD engineer. Other common career progressions for mudloggers/data engineers are as a wellsite geologist or drilling fluids engineer (mud engineer).

The complete list of responsibilities of the mudloggers is too exhaustive to detail in this article but the above-mentioned roles are the main ones. Like most jobs on the rig, daily reports are a big part of the data engineer's responsibilities.

The mudloggers report directly to the wellsite geologist, who is generally working in the mudlogging unit alongside them. Because the mudloggers are required to monitor the drilling operations from the commencement of drilling they will always be employed on a permanent rotating roster, which is generally 4-weeks on, 4-weeks off.

MWD / LWD Engineers and Directional Drillers

The terms Measurement While Drilling (MWD), and Logging While Drilling (LWD) are not used consistently throughout the industry. Although, these terms are related, the term MWD refers to directional-drilling measurements, while LWD refers to measurements concerning the geological formation made while drilling (also referred to as Formation Evaluation While Drilling (FEWD)).

Measurement While Drilling (MWD)

MWD typically concerns measurement taken of the wellbore inclination from vertical, and also magnetic direction from north. Using basic trigonometry, a three-dimensional plot of the path of the well can be produced.

Essentially, an MWD engineer measures the trajectory of the hole as it is drilled (for example, data updates arrive and are processed every few seconds or faster). This information is then used to drill in a pre-planned direction into the formation, which contains the oil, gas, water or condensate.

An MWD downhole tool is also "high-sided" with the bottom hole drilling assembly, enabling the wellbore to be steered in a chosen direction in 3D space known as directional drilling. Directional drillers rely on receiving accurate, quality tested data from the MWD engineer to allow them to keep the well safely on the planned trajectory.

MWD tools are generally capable of taking directional surveys in real time. The tool uses accelerometers and magnetometers to measure the inclination and azimuth of the wellbore at that location, and they then transmit that information to the surface.

With a series of surveys, measurements of inclination, azimuth, and tool face, at appropriate intervals (commonly every 30ft or 10m), the location of the wellbore can be calculated.

MWD tools can also provide information about the conditions at the drill bit. This may include:

- Rotational speed of the drillstring

- Smoothness of that rotation

- Type and severity of any vibration downhole

- Downhole temperature

- Torque and weight on bit, measured near the drill bit

- Mud flow volume

Use of this information can allow the operator to drill the well more efficiently, and to ensure that the MWD tool and any other downhole tools, such as a mud motor, rotary steerable systems, and LWD tools, are operated within their technical specifications to prevent tool failure. This information is also valuable to geologists responsible for the well information about the formation that is being drilled.

Logging While Drilling (LWD) tools and Formation Evaluation

The measurement of formation properties during the drilling of the hole through the use of tools integrated into the "bottom hole assembly" (BHA) can be expensive but has the advantage of measuring properties of a formation before drilling fluids invade deeply. Further, many wellbores prove to be difficult or even impossible to measure with conventional wireline tools, especially highly deviated wells. In these situations, the LWD measurement ensures that some measurement of the subsurface is captured in the event that wireline operations are not possible. Below is an example of an LWD/MWD bottom hole assembly.

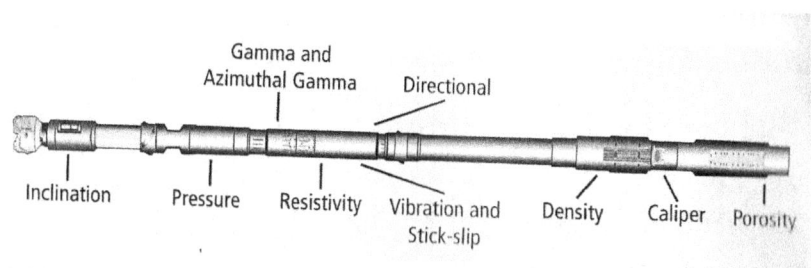

LWD tools take measurements of formation properties. At the surface, these measurements are assembled into a pictorial data log for fast and instant interpretation of the formation. LWD tools are able to measure a suite of geological characteristics including density, porosity, resistivity, acoustic-caliper, inclination at the drill bit (NBI), magnetic resonance and formation pressure.

The MWD tool allows these measurements to be taken and evaluated while the well is being drilled. This makes it possible to perform geosteering or directional drilling based on measured formation properties, rather than simply drilling into a preset target. Image logs are also possible, and there is an increase in demand for formation pressure tests and collection of fluid samples that can be obtained by increasingly sophisticated LWD tools.

Until recent years, pressure and fluid sampling could only be done when drilling was completed and wireline logs were run, but with the advances in LWD technology it is now becoming more routine to perform these tests while drilling the well so important drilling decisions can be made on the fly.

There are many different LWD tools available and every logging company has their own proprietary hardware and software. Tool mnemonics (acronyms used to explain the type of tool) feature heavily in formation evaluation programs as most logging tools, individual logging sensor measurements and log curves are known by their individual signature acronyms.

The three must-have curves you need for a basic well log analysis are: gamma ray, porosity and resistivity. These three curves give an excellent quick-look log analysis of reservoir formations and can give the wellsite geologist and shore-based petrophysicists an almost real-time preliminary interpretation of the zones of interest. This LWD tool data takes only seconds to get to surface and decoded into the data that is shown on the screens.

The time it takes from when the rock is drilled to when the data arrives at surface is dependent on how far behind the bit the individual LWD tools are positioned. For "Near Bit" tools, such as Gamma

Ray and Resistivity, this can be less than a metre so the information is received very soon after drilling.

Compare this data with how long it takes to get the actual "cuttings" to surface – which can be anywhere from 30 minutes to a couple of hours – and you can appreciate the benefits of having LWD tools in the BHA. In an operation where time equals money, you want your important decision-making data as soon as possible.

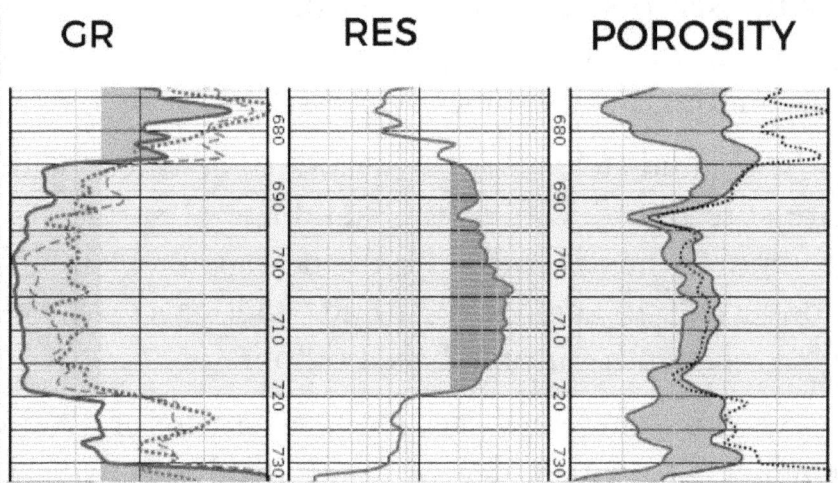

Data Acquisition

Downhole MWD hardware consists of sensors built

into a drill collar positioned near the bit. Electrical energy for the system is provided by a battery pack or generated by a small turbine. In a battery pack MWD system, information is recorded and stored downhole in the microprocessor. The data are retrieved when the MWD collar is brought to the surface and are transferred to the computer in the logging unit for additional processing.

In a typical turbine-powered "real time" MWD system, data are sent directly to the surface by *mud telemetry*, which utilizes the column of fluid inside the drill pipe as a transmission line for digital acoustic signals. Downhole measurements recorded by the sensors are transmitted through the mud as positive or negative pressure pulses or as a continuous, fixed-frequency pressure wave.

The mud telemetry signals are detected with pressure transducers in the standpipe at surface. A computer then records the digital signals. Data are converted to engineering units and processed to generate depth- or time-based output.

As you can see, it is almost essential for LWD engineers to have a degree in electrical, mechanical, chemical, petroleum or civil engineering although many also progress into it from mudlogging/data engineering positions within the same service provider company.

Because of the sometimes extreme physical

environment the tools are subjected to downhole (extremes of temperatures and pressures), tool wear and breakdowns are an all-too-common occurrence in LWD operations which places extreme pressure on the LWD operators to perform their job.

You have to be able to work under pressure and be thick skinned to be able to handle not only the troubleshooting operations but also the barrage of verbal abuse the LWD engineers are likely to face when their tools fail and "non productive time" is logged on the daily reports. If you think that character of the grumpy company man that John Malcovich portrayed in the "Deepwater Horizon" movie was exaggerated, think again…there really are people like that working on offshore rigs!

Wireline Logging - Formation Evaluation after Drilling

Wireline logs are recorded when the drilling tools are no longer in the hole and are made using highly specialized equipment entirely separate from that used while drilling. To run wireline logs, the hole is cleaned and stabilized and the drilling equipment pulled out of the hole. There is usually several wireline "runs" with different tools being used for different types of petrophysical data collection and formation sampling in each run.

After the well has been prepared for logging operations, the first logging tool is attached to the logging cable (wireline) and lowered into the hole to its maximum drilled depth. Most logs are run while pulling the tool up from the bottom of the hole, although just to be sure of having a record, measurements are recorded on the way down as well. The cable attached to the tool is both a support for the tool and a canal for data transmission and is wound around a motorized drum during the logging.

There is an instantaneous display as a log is acquired both on the rig and, if requested, by satellite link at the client's or operators shore-based office. Data is also stored electronically for future processing and editing.

Because rig time is expensive and holes must be logged immediately, modern logging tools are multi-function and multi-modular. Despite the use of combined tools, the recording of a full set of wireline logs still requires several different tool descents. While a quick shallow logging job may only take 3 – 4 hours, a deep-hole, full set may take

2 – 3 days or longer. Formation pressure testing and sampling runs can take up to 12 hours to perform each run.

The wireline operations are performed from the wireline unit, which is placed within close proximity to the drillfloor. The tools are lowered down the hole via a series of pulleys (sheaves) that direct the wireline cable from the drum at the unit to the open hole on the drillfloor. The wireline technicians will assist the wireline engineers with the pre-run tool checks, rigging up and rigging down of the tools and general maintenance of equipment while the engineers sit in the unit and monitor the data acquisition and processing during and after the run.

The wireline unit is extremely small and during wireline operations there will be at least three people working in the unit: the wireline engineer, the wireline technician operating the cable drum and the wellsite geologist.

The running of the logs is a very intense operation and constant monitoring by the wellsite geologist and wireline engineer is essential. It is quite often the case that with many of the logs (e.g. formation pressure testing and side wall core operations) both the wellsite geologist and wireline engineer don't get to leave the unit for their entire 12-hour shift – except for emergency bathroom breaks! Other crew members bring meals into them so they can eat while they continue to work.

Wireline engineers are sourced from the same educational backgrounds as MWD/LWD engineers. Electrical engineering is the most ideal base to be starting from but not the only route to get there. The wireline technicians don't necessarily need any formal qualifications as all their training is done on the job.

With technological advances in LWD tools and practices, wireline logging is slowly losing dominance as the main source of formation evaluation data. More and more services are being provided by LWD tools that mean many wireline contingent runs are no longer required. Despite this, wireline logging is, and will remain to be, a critical and necessary part of the offshore drilling operations.

When it comes to offshore mudlogging, MWD/LWD and wireline operations, the three major logging companies that service the industry are: Schlumberger, Halliburton and Baker Hughes. These three companies all have their own divisions of the individual services and some of them still operate under a name of a previous company that has been bought out by one of the big three mentioned (e.g. Geoservices mudlogging services). These three companies all have extensive shore-based support teams that work alongside the client petrophysicists and drilling team to provide timely and reliable data.

Chapter 9 of Offshore Oil and Gas PEOPLE will continue to explore the many different roles that are performed by third-party service providers on the rig.

CHAPTER 9

THIRD-PARTY SERVICE PROVIDERS
MISCELLANEOUS SPECIALTIES

Offshore oil and gas drilling rigs are a melting pot of races, cultures, professions and 21st century technology. While the general perception is that of grease-covered workers throwing tongs around the drillfloor, the reality is much different.

With every minute of the day having to be accounted for in daily reports and converted into monetary costs, it's no wonder only highly trained specialists are employed to undertake the myriad of roles performed in offshore drilling operations. Everyone on board the vessel works as a team to support the drilling operations and to make sure the well is drilled safely, and on time. Rig operations can cost up to (or even more than) $1 million dollars a day, which breaks down to up to $1,000 per minute for every minute of the 24-hour operations. This means that every minute of the day has to be accounted for and non-productive time (NPT) is not an option – well, it can be an option but you'll have a lot of explaining and arse-covering to do!

Routine testing and preventative maintenance are a huge component of the tasks performed by all offshore drilling contractors because when the time comes for their equipment to be used in the drilling operation they can't be causing delays by not having fully operational equipment.

Before any equipment goes "down the hole" it has to be fully tested, strapped (external dimensions measured) and drifted (internal measurements measured) to ensure compliance with very strict operational tolerances. Errors in calculations or faulty equipment can cost millions of dollars in lost productive time. Getting something wrong can see you with a one-way ticket on the next chopper!

I want to highlight a major difference between the salaried rig crew (although in the downturn this is now also true of the rig crew) and the third-party service providers (contractors). The contractors are regarded as dispensable – if you stuff up, you're generally out on your first strike. There's no soft-touch HR department on the rig that holds your hand and says: "Ohhh, we'll give you another chance"…if you want that treatment then you're in the wrong place! Go back to a cruisy 9 to 5 job in the city where managers aren't allowed to hurt your feelings…you're not going to get that out here.

And if you can't work 12-hour shifts for 28 days straight then you're also in big trouble. Twelve hours are a MINIMUM shift; during critical times of the drilling operations it's common to be "on-tour" for up to 15 hours or longer (with written approval) and if things are busy/bad enough that you have to do overtime then you can bet you won't have time for meal breaks during that shift either.

If the work schedule isn't enough to put you off, then be aware that the contractors are also given the shittiest rooms on the rig, which may even mean sharing a 4-man room with people who all work different shifts so your sleep gets disturbed every time someone enters the room.

It pays to learn to sleep with ear plugs in your ears because the shittiest rooms are always on the lowest level in the accommodation block and generally positioned over the pump room, the engine room or the anchor chain winches...or a combination of all three because on a small rig there's no escaping all of these! I'm not going to sugarcoat the jobs out there – the work and lifestyle can be tough and new-starters need to know this before they embark on a career offshore.

I'll go through the most common contractor jobs that are performed on the rig but there are many others that I won't have time to mention in this series of articles. These are the main ones that are pretty well always a part of the standard operations.

Remotely Operated Vehicle (ROV) Operators

ROVs are tethered, unoccupied, highly maneuverable remotely operated underwater vehicles, which are operated by a crew aboard the rig. They are linked to the rig by a load-carrying umbilical cable, which is used along with a tether management system (TMS). The TMS is either a garage-like device, which contains the ROV during lowering through the splash zone or, on larger work-class ROVs, a separate assembly that sits on top of the ROV. The purpose of the TMS is to lengthen and shorten the tether so the effect of cable drag where there are underwater currents is minimized. The umbilical cable is an armored cable that contains a group of electrical conductors and fiber optics that carry electric power, video, and data signals between the operator and the TMS. The TMS then relays the signals and power for the ROV down the tether cable.

Once at the ROV, the electric power is distributed between the components of the ROV. However, in high-power applications, most of the electric power drives a high-power electric motor, which drives a hydraulic pump. The pump is then used for propulsion and to power equipment such as torque tools and manipulator arms where electric motors would be too difficult to implement subsea. Most ROVs are equipped with at least a video camera and lights.

Additional equipment is commonly added to expand the vehicle's capabilities. These may include sonars, magnetometers, a manipulator or cutting arm, water samplers, and instruments that measure water clarity, water temperature, water density, sound velocity, light penetration, and temperature.

The ROV has a wide range of capabilities that include:

- Pre-spud surveying of the seabed and

underwater acoustic positioning systems

- Overseeing conductor cementing and wellhead installation operations

- Assisting with landing of the BOP on the wellhead

- BOP intervention by acting as an additional safeguard for operating emergency well shut-in procedures via the Blowout Preventer Actuating Tool (BOP-AT)

- Daily surveys to establish the integrity of all subsea components of the drilling operations, namely the riser, LMRP, BOP and wellhead.

- Daily checks of "bullseyes" on the wellhead and BOP to ensure there is no displacement from vertical of the seabed structures.

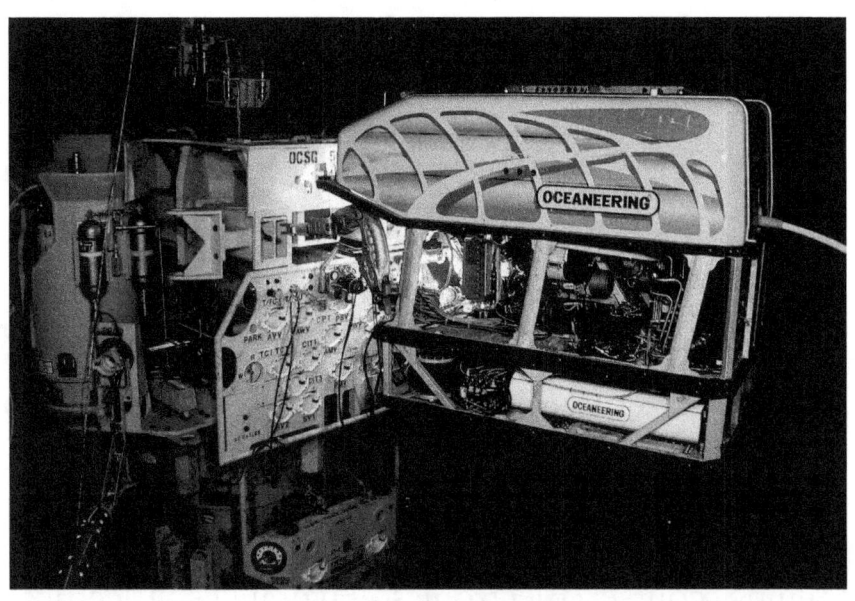

ROVs are commonplace on all rigs these days, especially in deepwater drilling. With exploration moving into water depths beyond that which can be usefully achieved by divers, the ROV becomes an increasingly important tool. Almost all ROV personnel are employed directly by ROV operators or contractors. There are a number of large internationally based ROV survey companies as well as many smaller operations.

A team comprising ROV pilots and technicians operates the ROV units. While both positions require certification in Hyperbaric Operations specific to operating ROV's offshore, flying an

ROV competently is not the only skill required of a pilot. ROVs are highly complex mechatronic (robotic) devices and working offshore in remote locations means assistance by qualified factory trained technicians is unavailable.

It is advantageous for the ROV pilot to hold appropriate technical and practical skills because they will have to be responsible for onsite repair and maintenance of the ROV unit under their care.

A qualification in one (or more) of the following nationally recognized trade skills (with post training employment) are considered to be essential pre-requisite qualifications for entry into an ROV training course: Electronics, Hydraulics, Electrical and Mechanical. Tertiary qualifications in an appropriate discipline, and significant relevant industrial experience may also help your chances of securing a spot in an offshore ROV team once your certification course has been completed.

The ROV team consists of at least two, but quite commonly three, personnel on each dayshift and nightshift. They will work out of a unit on the deck close to where the ROV docking station is located.

The atmosphere inside the ROV unit is quite unlike any other office on the rig, with serene underwater vistas showing on all the computer screens. In shallow waters there can be quite an aquarium affect showing on their monitors, especially in tropical waters. Added to this, there's usually a coffee-making machine, stereo and mood lighting – all essential elements of the ROV experience!

Solids Control

Solids control technicians are responsible for monitoring, operating and maintaining the centrifuge units that separate the solid cuttings and clay particles from the mud after the drilling fluid is circulated to surface.

The mud is piped via the "flowline" to the shale shakers, which are mesh-covered vibrating screens that sieve the solids from the liquid drilling fluid.

The separated mud is returned directly into the active mud system and circulated back down the hole while the solids (rock cuttings) are directed into a rotating auger under the front of the shaker screens, which then flows into the centrifuge units.

If the ROP (rate of penetration) is very high then there can be a high volume of cuttings coming over the shakers and the centrifuge units can have trouble keeping up with the demand. If this happens then the units can stall and halt the flow of cuttings that can be treated.

This has to be monitored very carefully because any disruption to the flow of solids coming out of the hole can have a huge knock-on effect in many areas, including the possibility of an environmental incident if centrifuged cuttings are being discharged into the sea.

In some regions of the world the discarded centrifuged cuttings will be allowed to be discharged overboard into the sea but there are strict regulations regarding the amount of residual mud they can contain. The operating company's Environmental Impact Study will detail this and it must be strictly adhered to. In other regions this practice is forbidden so the discarded cuttings need to be stored in skips, which get back loaded to a dumping location onshore.

Centrifuge technicians will generally have a mechanical trade background as their primary function is to keep the machines working. If the centrifuge units breakdown then they have to be fixed onsite, if possible. Any delays in repairs can cost valuable drilling time…and remember: NPT (non-productive time) is not your friend!

Casing and Cementing Operators

While the drill crew is in charge of drilling the actual

well, specialist casing and cementing crews perform the running and cementing of the casing strings after each section has been drilled.

The well is drilled in stages whereby it is drilled to a certain depth, cased and cemented, and then the well is drilled to a deeper depth, cased and cemented again, and so on.

Each time the well is cased, a smaller diameter casing is used. The widest type of casing is called *conductor* pipe, and it is usually about 30 to 42 inches in diameter. The next size in casing string is the *surface* casing, which can run several thousand feet in length.

Intermediate casing is then run to separate challenging areas or problem zones, including areas of high pressure or lost circulation. The last type of casing string that is run into the well, and therefore the smallest in diameter, is the *production* casing, which is run directly into the expected reservoir zone.

In an effort to save money, sometimes a *liner string* is run into the well instead of a casing string. While a liner string is very similar to casing string in that it is made up of separate joints of tubing, the liner string is not run the complete length of the well.

A liner string is hung in the well by a liner hanger, and then cemented into place.

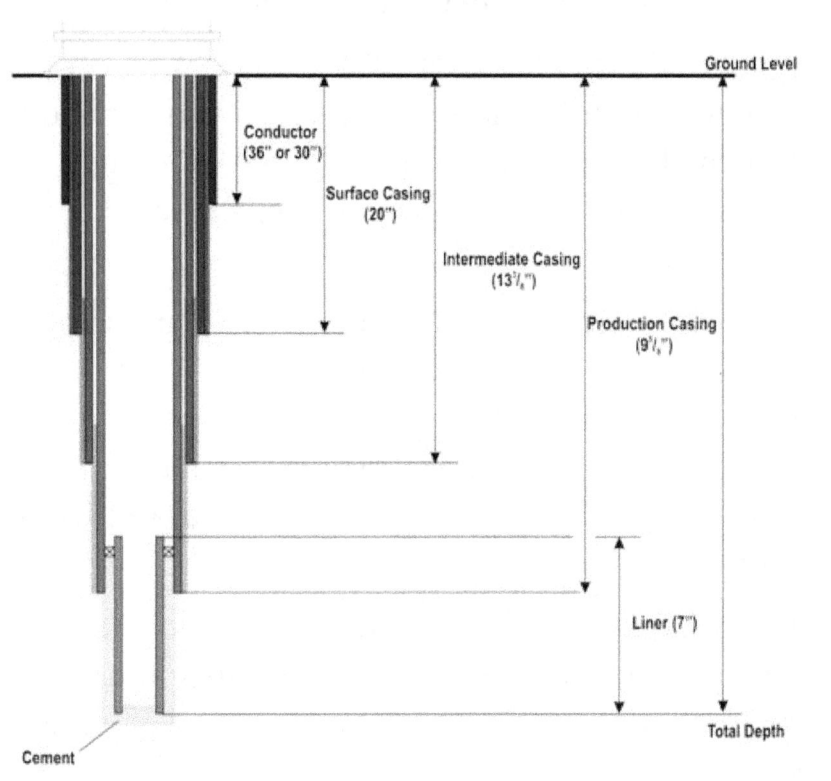

Casing is run from the rig floor, connected one joint at a time by casing elevators on the traveling block and stabbed into the previous casing string that has been inserted into the well. Hanging above the drill floor, casing tongs screw each casing joint to the casing string.

Casing is run into the well and officially landed

when the weight of the casing string is transferred to the casing hangers, which are located at the top of the well and use slips or threads to suspend the casing in the well.

A rounded section of pipe with an open hole on the end, a *guide shoe* is connected to the first casing string to guide the casing crew in running the casing into the well. Additionally, the outside of the casing has spring-like *centralizers* attached to them to help position the casing string in the center of the well.

A cement slurry is then pumped into the well and allowed to harden to permanently fix the casing in place. After the cement has hardened, the bottom of the well is drilled out, and the completion process continues.

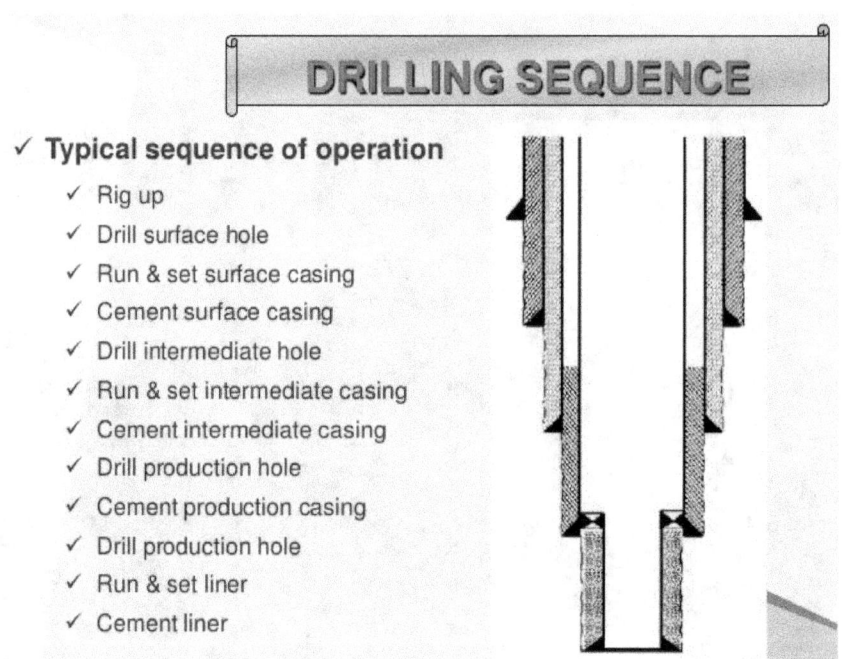

DRILLING SEQUENCE

✓ **Typical sequence of operation**
  ✓ Rig up
  ✓ Drill surface hole
  ✓ Run & set surface casing
  ✓ Cement surface casing
  ✓ Drill intermediate hole
  ✓ Run & set intermediate casing
  ✓ Cement intermediate casing
  ✓ Drill production hole
  ✓ Cement production casing
  ✓ Drill production hole
  ✓ Run & set liner
  ✓ Cement liner

While the casing hands and cementers receive assistance from the drill crew when needed, these specialists will oversee all the technical aspects of the job.

As you may have seen in the "Deepwater Horizon" movie, the cementing jobs are a critical part of the process safety systems of the well and the integrity of the cement bond is crucial to the safe operation of all the procedures that follow. This step in the drilling program is so critical that the company man will personally oversee the operation every step of the way to ensure total compliance with procedures and expected outcomes.

The lessons learnt from the Deepwater Horizon incident show the importance of overlooking psychological bias when interpreting cement integrity well tests. The investigation into this disaster highlighted how easy it is to skew the evidence in favour of what you want the outcome to be, despite evidence indicating otherwise.

The cementing and casing programs on all offshore wells are highly technical and very detailed. They are created through a collaboration of onshore teams in the drilling department and the specialist companies that are contracted to provide the service. Every step of the process is captured in computer monitoring systems on the rig and this data is interrogated by both offshore and onshore personnel to ensure the strict procedures set out in the well plan are being followed.

Detailed well plans are drawn up in advance of the drilling of wells being commenced and they have to be signed off and approved by many departments and regulatory departments. By law, these plans have to be followed exactly as detailed, as they will be based on "best practice" and strict safety requirements.

If for any reason these plans need to be changed, based on changes encountered during the drilling process that were not considered when preparing the well plan, then there needs to be a "Management of Change" document prepared. This "MoC" then needs to get approval from the highest levels in both the oil and gas companies and the third-party service providers so everyone agrees that the changes can be made without compromising the safety of the well. The cost in delays while waiting for approval can quickly add up to the millions of dollars so any deviation from "the plan" places a lot of stress on the people involved in the operation.

One last point that should be made about the casing hands and cementers is that they are advised to take out a good supply of books or movies to the rig with them. As both jobs are only performed at certain times during the drilling process, both crews are on stand-by on the rig waiting for their turn to shine. If the drilling process suffers no setbacks or delays then the casing and cementing operations will be quite fast-paced. However, quite often there *are* setbacks or delays that can take days, or even weeks, to resolve so the casing and cementing personnel have nothing more to do but check, and recheck, their equipment for all of this time.

Well Testing Engineers

If hydrocarbon zones are intersected while drilling an offshore well there is likely to be a series of well tests carried out to evaluate the reservoir potential. The overall objective is to identify the reservoir's capacity to produce hydrocarbons.

Test objectives will change throughout the different phases of a reservoir or oil field, from the exploration phase of wildcat and appraisal wells, through the field development phase and finally through the production phase, which may also have variations from the initial period of production to improved recovery by the end of the field lifecycle time.

The main objective in the exploration phase is to assess the size of a reservoir and state with a given certainty whether it has the properties for commercial exploitation and shall contribute to accounting for available reserves.

Well testing taking place before permanent well

completion is referred to as drill stem testing (DST).

A DST is a procedure for isolating and testing the pressure, permeability and productive capacity of a geological formation during the drilling of a well. The test is an important measurement of pressure behavior at the drill stem and is a valuable way of obtaining information on the formation fluid and establishing whether a well has found a commercial hydrocarbon reservoir.

In a drill stem test, the drill bit is removed and replaced with the DST tool and devices are inflated above and below the section to be tested. These devices are known as packers and are used to make a seal between the borehole wall and the drill pipe, isolating the region of interest. A valve is opened, reducing the pressure in the drill stem to surface pressure, causing fluid to flow out of the packed-off formation and up to the surface.

There are two distinct phases of the DST's and that is the "flowing" phase and the "shut in" phase.

During the flowing phase in exploration and appraisal wells the following information is gathered:

- Confirmation of discovery and productivity

- Volumetric flow behavior and rate

- Clean-up and rate measurement

- Hydrocarbon properties and characteristics of the reservoir

- Pressure

- Gas oil ratio

- Collection of large volume fluid samples both down hole and at the surface

- Testing of sand production

During the "shut-in" phase, which will commonly contain at least two "pressure build up" tests, the following data can be established:

- Well and reservoir performance (skin, permeability, initial pressure, heterogeneity and boundaries)

- Reservoir connectivity and proven volume

- Flow behavior around the well bore

The following YouTube videos demonstrate the DST procedure both in open holes and cased holes. I am in no way endorsing Expro here but they had a very good video explaining the process, for which they deserve some credit.

DST open hole testing:
https://www.youtube.com/watch?v=xpuiyXD
YUkY

DST cased hole testing:
https://www.youtube.com/watch?v=saWFV26
UmcQ

Due to the hydrocarbons being brought to the surface there will always be a flare burning during testing operations. The flare boom will either be off to the side or the back of the rig with water deluge systems suppressing the enormous amount of heat this flare generates.

Well testing engineers and technicians come from a

varied background of experience in mechanical, electrical, petroleum or reservoir engineering or trade skills.

Like many of the third party service providers, well testing technicians often start off in the offshore industry as roustabouts and then train in specialty fields.

Well testing crews don't generally have a permanent rotation on a rig because they are only needed at the end of the drilling of the wells so they are flown in only as they are needed. They will be scheduled to arrive on the rig several days prior to the expected completion of drilling so they can set up and test their gear before starting the DST operations.

Flaring always presents a great photo opportunity for rig workers who are starved of allowable larrikin antics.

While there are many other third-party service providers that work in offshore drilling operations it's impossible to cover them all within the scope of these articles. The ones I've mentioned are the most common and generally always present in all drilling operations.

There's just one more key group of contractors that work on the rig that's of vital importance and that is the accommodation and catering crew. While they are third-party contractors they are actually employed by the drilling contractor and as such, they report directly to the OIM.

Catering Contractors (Stewards)

For people working offshore, a hot meal, a clean bed and freshly laundered work wear are essential comforts. With everyone working a minimum of 12-hour shifts every day it takes a very well structured support system to make sure everyone is adequately fed and has a clean room to sleep in at the end of the working day.

Some drilling contractors employ their own catering crew while many outsource the tasks to third-party catering companies. The standard that is expected is to have all cabins and bathrooms serviced daily, which nearly always includes having your bed made for you, and clean towels provided. Because of the lack of living space, and strict baggage restrictions for all personnel flying to a rig, everybody's clothes are laundered daily to minimize the amount of clothing required. During the boom times there were usually added extras like lifestyle coaches and personal trainers making regular visits to the rig but that service disappeared once the price of oil started to drop.

The quality of meals is extremely variable, depending on the rig and location around the world. You can expect to have the cuisine of whatever country you are drilling in so this can vary from

American food if you're in the Gulf of Mexico, British food if you're in the North Sea, Indian food if you are in many SE Asian regions, or any number of other variations. Being unskilled laborers, the catering and cleaning crew will be sourced from the closest port to where the rig is drilling.

With crew changes occurring on a daily basis, and bed space usually filled to capacity, it is necessary to have dayshift and nightshift cleaners, as well as cooks. These are generally entry-level jobs that require no experience although many of the people have experience in similar roles at onshore mining camps or similarly serviced remote work sites. While the large international contracting service providers tend to have a gender-balanced workforce these days, it's still rare (in my experience) to see any women working these roles on rigs that use their own catering and cleaning crews. Some habits are hard to break!

While on the gender issue, it's probably worth noting that there are generally very few women working on offshore drilling rigs. The roles that are most likely to have females represented are catering, mudloggers and MWD. I have, on many occasions, been the only female on the rig out of a POB of up to 180. While it's rare to be the ONLY woman on board, it does occur from time to time, but generally there will be a few scattered around the facility in different roles. You generally won't see more than half a dozen women working on a rig out of 120 to 180 workers on board. More than likely it will only be two or three.

****

With all the drilling contractor crews and third-party service providers now covered, it just leaves the oil and gas company's representatives to cover in the final chapter of Offshore Oil and Gas PEOPLE. Chapter 10 will explore the main roles carried out by these oil and gas professionals and what it takes to be part of the team.

CHAPTER 10

OPERATING COMPANY
REPRESENTATIVES

With this being the final chapter, I thought it a good idea to list all the roles that we cover in this book. This is by no means an exhaustive list of everyone you are likely to find working on an offshore drilling rig but they are the key roles that you will always find on every rig. There are many more specialists that work in offshore drilling operations but they are too numerous to mention in the scope of these articles.

DRILLING CONTRACTOR			
Deck Crew	**Drill Crew**	**Subsea**	**Marine Crew**
Roustabouts	Roughnecks	Subsea Engineers	Captain/Master
Crane operators	Derrickman	Subsea Technicians	Chief Engineer
Deck Pusher	Assistant driller		Dynamic Positioning Operators
	Driller		Radio Operators
	Toolpusher		Ballast Control Operators
	Offshore Installation Manager		Mechanics and Electricians
			Medics
			Materials Coordinator
			RSTC
THIRD PARTY SERVICE PROVIDERS			
Formation Evaluation and Well Monitoring		**Other Service Providers**	
Sample Catchers		ROV Operators	
Mudloggers		Casing Hands	
Data Engineers		Cementers	
MWD/LWD Engineers		Well Testing Engineers and Technicians	
Wireline Engineers and Technicians		Catering - Stewards	
OIL AND GAS COMPANY REPRESENTATIVES			
Company Man (Wellsite Manager)		Drilling Fluids (Mud) Engineer	
Drilling Engineer		Wellsite Geologist	
Logistics Coordinator		HSE Coordinator	

This chapter will cover the key personnel who are the operating oil and gas company's representatives on the rig during drilling operations. Some may be salaried employees of the energy company while others will be contracted specialists who report directly to the operating company.

We'll start with the head honcho of the offshore operations, the Company Man (Wellsite Manager), and work our way down the pay scale – which is not an indication of level of responsibility or workload but rather the higher degree of specialization required to perform the role.

Please don't quote me on this order though, because pay scales differ considerably from one contract to another and depending on boom or bust times in the industry. Some of these people (if they are lucky enough to still have a job) would possibly be on 50% less now (February 2017) than what they were getting paid in February 2015.

Company Man / Wellsite Manager (WSM)

The company man, or more commonly used these days is the title wellsite manager (WSM), is the point of contact on the rig for the operating company. He has absolute responsibility over all personnel, financial, technical and performance aspects of the drilling program and the rig being contracted to carry it out. Nothing is supposed to get done on the rig without the company man knowing about it.

If any well control barriers are compromised in any way, he needs to know about it immediately. It's pretty fair to assume that whenever you hear the company man being paged over the rigs PA system by a supervisor, shit has probably just hit the fan somewhere on the rig!

All major drilling decisions moving forward of any point in the program will generally be made in consultation with the drilling superintendent in the head office onshore and any specialists required, but ultimately any immediate action responses are the responsibility of the company man.

There will always be a dayshift company man and nightshift company man with the day company man being the most senior of the two. If there are any major changes to the plan, or emergency situations arise overnight then the night company man will wake the day company man to brief him on what has happened and get advice on how he wants the situation to be handled.

While the day WSM will spend most of his shift sitting in front of a computer or on conference calls in the office, the night WSM will spend more time on the drillfloor or in the doghouse overseeing the drilling operations. One of the two will always be present on the drillfloor during critical well control operations and also during cementing operations because of the process safety implications of the procedure.

Wellsite managers come from a variety of

backgrounds but one thing is for sure – they have usually all lived and breathed offshore oilfield all their lives and have done the hard yards to get where they are.

The two most common career paths are via the roughneck-driller-toolpusher route or the drilling engineer route. Occasionally you'll come across a WSM that started out as a geologist or some other profession but it's more likely they used to be a driller or a drilling engineer.

The longer and harder their career progression then the more respect they generally earn from the crew under their command. A "young" ex-drilling engineer who has never worked outside an air-conditioned office will never earn quite the same level of respect as that given to an old driller who has spent 30-40 years of his life "on the tools" and sweating it out over decades of hard yakka on a rig.

The "day" WSM will nearly always have had to earn his stripes by first working as a "night" WSM. They are two very distinct positions. With nearly every other role on a rig the dayshift and nightshift crew interchange with each other but this is generally not the case with the company man, except in unusual circumstances.

The day WSM will quite often be a salaried

employee of the operating company and be heavily involved in the planning and preparation stages of the drilling campaign back in the office, long before a rig is ever mobilised to the drilling location. He needs to know every stage of the drilling plan and know what all the other workers on the rig are meant to be doing at all stages of the operation.

Many company men would have worked a wide variety of offshore drilling roles throughout their career and be very familiar with many of the specialist operations. The night guy will commonly be a contracted worker.

The day WSM is one of the few people who generally get their own cabin. The night WSM sometimes has his own room also but more often would share with the logistics guy, or someone else who only works dayshift. With space being a premium on any rig, all 2-man cabins generally house a dayshift person and a nightshift person so they are always likely to have a room to themselves while off-tour.

Drilling Engineer

The drilling engineer is the company man's eyes and ears on the rig. He will generally spend a lot of time in the doghouse with the driller and toolpusher and physically oversee all aspects of the well operations so he can report back to the WSM.

There is generally only one drilling engineer on the rig at any one time, working on dayshift (0600-1800hrs) unless operations otherwise dictate.

Other duties the Drilling Engineer is responsible for include (but definitely not limited to):

- Closely monitoring day-to-day operations and reporting back all findings and observations to the WSM.

- Collecting and analyzing data relating to all stages of the drilling operations.

- Performing, recording and disseminating results of "after action reviews" after all key stages of the operation.

- Keeping track of all costs accrued through the drilling of the well.

- Working closely with all on-site specialists in order to keep up-to-date on all developments that may have an impact on drilling activities.

- Making sure that drilling operations comply with statutory and regulatory requirements, with respect to health and safety, emergency procedures and disaster recovery.

- Continually revising and updating the forward plan and projected timings of the well operations so the logistics coordinator and third-party service providers can accurately prepare equipment in a timely manner.

Drilling engineers will generally have completed some sort of engineering degree prior to being employed by an oil and gas company.

Initial training and development is primarily facilitated through graduate development schemes, which involve gaining hands-on work experience through multiple rotations offshore and in-house training sessions.

Career progression is mainly driven by individual performance, professional expertise and attainment of professional qualifications.

Wellsite Geologist (WSG)

The wellsite geologist is the source of all operational geological information on the rig and is responsible for all geology related administrative wellsite activity. They are the exploration department's eyes and ears on the rig and as such, have to make sure that all possible geological and drilling information is gathered in a concise and timely manner.

While the WSG works in close cooperation with the company man on the rig he is not actually under his authority. Instead, the WSG reports directly to the "Operations Geologist" who is the "shore based" intermediary between the geologist on the rig and the geology team in town who will be analyzing all the data. The unusual chain of command for disseminating key official geological data from the wellsite geologist follows this line of reporting:

WSG (rig) => Operations Geologist (town) => Drilling Superintendent (town) => Company Man (rig)

While the wellsite geologist is required to immediately notify the company man of any pertinent drilling and geological information, the company man generally cannot act on the information until the town-based drilling superintendent has officially confirmed it.

The WSG will report all key geological and drilling data to the operations geologist immediately as it comes to hand. It is then the responsibility of the "ops geo" to disseminate this information to all members of the onshore geology and drilling teams who need to know the information for decision-making.

All key drilling decisions are made in collaboration with every department involved in the drilling of the well to ensure that well control barrier criteria are met and any decisions made will not compromise the integrity of the well or process safety systems.

At commencement of drilling, when the well will be drilled "riserless" with no cuttings coming to surface, there will often only be one WSG on the rig. There may be two or even three casing strings run before the riser is finally run and drilled cuttings are brought to the surface.

The WSG will be needed during these stages of drilling to confirm that suitable geological formations have been intersected in order to successfully set casing. This task is commonly referred to as "calling casing point". It is critical that the casing shoe for the conductor and surface casing is set deep enough to withstand pressure from a "kicking" formation further down.

Surface casing is run to prevent caving of weak formations that are encountered at shallow depths. The WSG needs to identify when a competent formation is intersected to ensure that the formation at the casing shoe will not fracture at high hydrostatic pressure, which may be encountered later in the drilling of the well.

Because there are no drilled cuttings coming to surface all geological data is interpreted from one, or a combination of both, of the following sources:

- Drilling parameters such as ROP and torque when there are no LWD (Logging While Drilling) tools in the BHA (Bottom Hole Assembly).
- Real time Gamma Ray and/or Resistivity data from downhole LWD tools.

Once the surface casing has been set and the BOP and riser are run to the seabed, all drilled cuttings will then be circulated to the surface, which means the days get a whole lot busier for the WSG. From this stage on there will generally be two WSG's operating back-to-back 12-hour shifts.

Responsibilities

As the acting representative for the operating company's geology team, the wellsite geologist will have the following responsibilities:

- Evaluating offset data before the start of drilling
- Analyzing, evaluating and describing formations while drilling, using cuttings, gas, formation evaluation measurement while

drilling (FEMWD) and wireline data

- Comparing data gathered during drilling with predictions made at the exploration stage;

- Advising on drilling hazards and drilling bit optimization

- Making decisions about suspending or continuing drilling. Ultimately, it's the wellsite geologist's responsibility to decide when drilling should be suspended or stopped.

- Advising operations personnel both on the rig and in the onshore operations office about any pertinent geological or drilling information as it arises.

- Supervising mudlogging, MWD/LWD and wireline services personnel and monitoring quality control in relation to these services.

- Keeping detailed records, writing reports, completing daily, weekly and post-well reporting logs and sending these to appropriate departments.

- Maintaining up-to-date knowledge of LWD and MWD tools and status of all equipment onboard and in transit to make sure the equipment is available and in working order when it is needed.

In expected HPHT (high pressure high temperature) wells it is critical the WSG can identify (and immediately communicate) any identifying signs of increases in pore pressure. These can include the following telltale signs:

- Changes in flow rate and active mud system volumes. If the formation pressure becomes higher than the hydrostatic pressure being exerted by the circulating drilling fluid then the mud will become "underbalanced" and the well will "kick". If this kick isn't detected early enough then a catastrophic blowout could occur.

- Presence of "cavings" coming over the shakers. When drilling over-pressured shales, it is common for the formation to undergo stress relief causing chips of rocks to cave from the borehole wall. These overpressure "cavings" tend to be larger than normal cuttings and may be concave or propeller shaped.

- Increase in ROP (rate of penetration) and volume of cuttings. A pressure transition zone will make drilling easier because of the trapped water reducing compaction and the increase in pore pressure reducing differential pressure, allowing cuttings to be released more easily into the mud stream.

- Changes in LWD data, in particular resistivity and sonic, density and neutron.

- Changes in drilling parameters, especially torque, drag and overpull. This can be due to deterioration of borehole integrity causing an increase in volume of cuttings and cavings in the circulating mud.

- Rise in background gas level, changes in the composition of the gas, or presence of "connection" gas, which is a result of swabbing downhole hole when the pumps are turned off to make a connection (add another stand of drillpipe).

- Changes in pump pressure. An influx of gas into a well may reduce the density of the drilling fluid and therefore it will require less pressure to circulate the drilling fluid.

- Change in properties of mud.

- Changes in downhole temperature. Generally there will be slight decrease in temperature immediately above the over-pressured zone and then a steady increase with depth at a higher rate than in the normally pressured zone above.

If the wellsite geologist identifies any potentially hazardous changes in the drilling, the driller and company man must be notified immediately, and then the operations geologist will be notified.

If a potentially dangerous situation is recognized then the drilling will be stopped immediately while the company man either makes a decision on what to do next or waits for official instructions from the drilling superintendent in town on how to proceed.

The wellsite geologists spend most of their time working in the mudlogging unit (like the hardworking one in the photo above ☺), which is where all the monitoring equipment for the rig is located and also where the mudloggers/sample catchers will deliver the cuttings samples for them to inspect and describe.

All rock cuttings are inspected under a microscope and a detailed description written for every sample that is generally collected in composite 5, 10 or 20 m intervals.

Cuttings Descriptions

The cuttings descriptions need to be very detailed and follow an industry standard format that includes (but is not restricted to) the following observations:

- Rock types and percentage of each found in the sample
- Color
- Texture
- Grain or crystal size
- Sphericity, roundness and sorting of sandstone grains
- Type of cement and/or matrix
- Any fossils or accessory minerals
- Presence of hydrocarbon indications, such as

fluorescence or "show"

- Estimate of porosity

Red Shale

Brown Shale

Gray Shale

Limestone

Sand grain

Consolidated sand

Limestone conglomerate
w/ imbedded glauconite (green)

Sample of drill cuttings under a 10x microscope

A detailed well log is created combining all the cuttings information, LWD and MWD data and drilling parameter data, and submitted along with a daily report every 24 hours. When the WSG finishes the shift and hands over to the next shift they have to have all of the reporting and samples descriptions up-to-date at the time of them handing over.

To become a wellsite geologist, you'll need a degree in geology or possibly even chemistry, geochemistry or geophysics. There is no formal wellsite geologist qualification, but you would need to obtain knowledge in areas such as wellsite and offshore safety management, wellsite operations, formation evaluation of wireline, FEWD logs, and risk assessment before starting as a WSG.

Most WSG's start their offshore career working as a mudlogger, MWD engineer or mud engineer and gain knowledge in the fields that a WSG is responsible for. They also need to possess supervisory skills, the ability to work well under pressure and the ability to quickly make decisions.

As most wellsite geologists work as independent consultants and are employed on a contracting basis, it's up to them to handle their own career progression. Any wellsite geologists who progress beyond this position will generally move into an operations geologist role, with a few even moving up into company man positions.

While a wellsite geologist might earn a lot per day there is little job security, and quite often no permanent rotation. They may only get flown onto the rig the day before drilling operations begin and flown off again immediately after the well is completed or wireline logging is completed. The date of your arrival and departure is quite often only known within days of it occurring so long term social commitments are impossible to plan. You can either expect to have to fly out to the rig at very short notice or have unplanned months without any work…or even years, as the case is for many now!

Drilling Fluids (Mud) Engineer

The drilling fluids engineer, who is most commonly referred to as the mud engineer, or just the mud man, is the person responsible for ensuring the drilling fluid properties are within designed specifications.

The drilling fluid (mud) is a vital part of drilling operations and has the following functions:

- Provides hydrostatic pressure on the borehole wall to prevent uncontrolled production of reservoir fluids.

- Lubricates and cools the drill bit

- Carries the drill cuttings up to the surface

- Forms a "filter-cake" on the borehole wall to prevent drilling fluid invasion into the formation

- Provides an information medium for well logging

- Helps the drilling by fracturing the rock from

the jets in the bit.

One of the most important mud properties is the mud weight (density). If the mud weight exceeds the fracture pressure of the formation, the formation may fracture and large quantities of mud can be lost to it, in a situation referred to as lost circulation.

If the mud weight is too low it will have a hydrostatic pressure that is less than the formation pressure. This will cause pressurized fluid in the formation to flow into the wellbore and make its way to the surface. This is referred to as a formation "kick" and can lead to a potentially deadly blowout if the invading fluid reaches the surface uncontrolled.

To maximize the effectiveness of these tasks, the mud contains carefully chosen additives to control its chemical and rheological properties.

For the technically minded, the drilling mud is usually a shear thinning non-Newtonian fluid of variable viscosity. When it is under more shear, such as in the pipe to the bit and through the bit nozzles, viscosity is lower which reduces pumping-power requirements. When returning to the surface through the annulus it is under less shear stress and becomes more viscous, and hence better able to carry the rock cuttings.

Bentonite is commonly used as an additive to control and maintain viscosity, and also has the additional benefit of forming a mud-cake (also known as a filter cake) on the borehole wall, preventing fluid invasion into the formation.

Barite is commonly used to increase the mud weight to maintain adequate hydrostatic pressure downhole in order to avoid a kick and ultimately a blowout from uncontrolled production of formation fluids.

The mud pits at the surface have their levels carefully monitored, since an increase in the mud level indicates a kick is taking place, and may require shutting in the well and circulating heavier weighted drilling mud to prevent further formation fluid or gas production.

The drilling mud must be chemically compatible with the formations being drilled; in particular the salinity must be chosen so as not to cause clay swelling or other problems. Offshore rigs typically use synthetic oil-based mud although water-based mud is also sometimes used.

Prior to drilling a well, a mud program will be worked out according to the expected geology, in which products to be used, concentrations of those products, and fluid specifications at different depths are all predetermined. As the hole is drilled and gets deeper, more mud is required, and the mud engineer is responsible for making sure that the new mud to be added is made up to the required specifications. The chemical composition of the mud will be designed so as to stabilize the hole.

As drilling proceeds, the mud engineer will get information from other service providers such as the mudlogger about progress through the geological zones, and will make regular physical and chemical checks on the drilling mud.

The viscosity and density are frequently checked. As drilling proceeds, the mud tends to accumulate small particles of the rocks that are being drilled through, and its properties change. It is the job of the mud engineer to specify additives to correct these changes, or to partially or wholly replace the mud when necessary.

Mud engineers come from a varied background, with many having no formal tertiary qualification but rather have had offshore drilling experience within one of the many other roles found on an offshore drilling rig. It's common to find mudloggers with a geology background transferring into the higher-paid role of the mud engineer.

Prior to working on his own, the junior mud engineer will have attended a special training course and will spend time working with a senior mud engineer to gain experience. The least experienced mud engineer will commonly work permanently on nightshift with the experienced mud man working days so he can communicate with the company man and onshore drilling support team.

The derrickman and roughnecks are assigned to help the mud man whenever he needs assistance with altering the mud properties or any other pit-related work. With the mud being one of the key process safety barriers in the drilling process the mud engineers are always kept busy monitoring it.

Drilling fluids operations are often contracted to service companies, with the largest four companies for mud services being M-I SWACO (A Schlumberger Company), Baroid Drilling Fluids (Halliburton Oilfield Services), Baker Hughes Drilling Fluids, and Weatherford International Drilling Fluids.

Health, Safety and Environmental Coordinator (HSEC)

Many offshore drilling operations will have a HSE representative within the drilling contractor crew (commonly referred to as the Rig Safety and Training Coordinator or RSTC) and also the operating company will have their own health and safety representatives. There may be a nightshift and dayshift or just the one person who works mostly dayshift, except when needed if there is a safety incident overnight. They can be either an employee of the operating oil and gas company or a contract worker.

The responsibility of the HSE coordinators is to ensure that all tasks on the rig are completed in accordance with company and regulatory requirements by using approved procedures and permits.

Safety reps normally come from quite varied backgrounds, with many having worked other roles within the drilling industry or sometimes even come from a military background.

A background as a rig worker is most advantageous because they would then have a competent knowledge of the tasks performed on the rig as well as all the equipment being used. This knowledge would make investigations and report writing of incidents a lot easier.

As well as the unofficial duties as the rig psychologist, auditor, mentor, deckhand, and personal problem advisor, the HSE coordinator also has to fulfill the following official tasks:

- Monitor the safe operation of all workers on the rig

- Participate in key project management activities e.g. HAZIDs and HAZOPs

- Provide management system documentation development and implementation

- Incident investigation

- OHS auditing

- Conduct weekly safety meetings and disseminate incident investigation findings from other areas within the industry

With the industry becoming increasingly heavily regulated the safety rep will be kept busy filling in paperwork and completing safety audits in between investigating incidents and report writing. He will also quite often work out of an office that is close to the company man's office *and* has a coffee machine, so it's the obvious place to kill time while waiting for the morning meetings to start.

Drilling Logistics Coordinator (DLC)

The last role to be covered in this book is the

drilling (and materials) logistics coordinator. The DLC is responsible for coordination of all materials, personnel movements and logistics support for the rigs operations. Key responsibilities are:

- Liaise with key personnel for timely provision of personnel, services, equipment and materials

- Liaise with key onshore supply operations personnel for the load-out and back-load of equipment and materials

- Coordinate storage, maintenance, record keeping and reporting of all the company's and contractor equipment on the rig.

The DLC will commonly work out of the same office as the company man, or close to it, so he can communicate all material and people movements and current state of operations in order to provide timely logistics advice to service providers and rig crew. There is generally only one onboard at any one time, working dayshift hours unless otherwise needed.

A background in rig operations through either working as a member of the drilling or deck crew, or as a service provider is advantageous, as they will need to know the name of, and be able to identify every part needed on the rig. Their backgrounds are generally quite varied, as it is not a position that requires a formal certification.

That concludes the

Offshore Oil and Gas PEOPLE

story, and while it was by no means an exhaustive list of roles performed in offshore drilling operations it covered the main ones. I hope you have a better understanding now of the main roles performed in offshore drilling operations and the people who carry them out.

ABOUT THE AUTHOR

Amanda Barlow is a contract geologist who has worked within the offshore oil and gas, coal seam gas and minerals industries – a career spanning over 30 years. She can be contacted through her LinkedIn profile.

https://www.linkedin.com/in/amanda-barlow-wsg

Amanda is also a recreational marathon runner and a published author of two other books. She has run 42 marathons in 16 different countries around the world, including the Jungle Marathon, in Brazil, the story of which was told in her first published book: "Call of the Jungle – How a Camping-Hating City-Slicker Mum Survived an Ultra Endurance race through the Amazon Jungle" (available in print and kindle version on Amazon.com).

Her second book: "An Inconvenient Life: My Unconventional Career as a Wellsite Geologist" documents her 30+ year career as a Geologist and how she ended up working as a wellsite geologist on offshore oil and gas rigs. It is also available in kindle and print versions on Amazon.com and all other eBook distributers, including iBooks.

You can follow "An Inconvenient Life" on Facebook at
https://www.facebook.com/AnInconvenientLife